Friction Based Additive Manufacturing Technologies
Principles for Building in Solid State, Benefits, Limitations, and Applications

Friction Based Additive Manufacturing Technologies
Principles for Building in Solid State, Benefits, Limitations, and Applications

Sandeep Rathee
Manu Srivastava
Sachin Maheshwari
T. K. Kundra
Arshad Noor Siddiquee

CRC Press is an imprint of the
Taylor & Francis Group, an **informa** business

CRC Press
Taylor & Francis Group
6000 Broken Sound Parkway NW, Suite 300
Boca Raton, FL 33487-2742

© 2018 by Taylor & Francis Group, LLC
CRC Press is an imprint of Taylor & Francis Group, an Informa business

No claim to original U.S. Government works

Printed on acid-free paper

International Standard Book Number-13: 978-0-8153-9236-1 (Hardback)

This book contains information obtained from authentic and highly regarded sources. Reasonable efforts have been made to publish reliable data and information, but the author and publisher cannot assume responsibility for the validity of all materials or the consequences of their use. The authors and publishers have attempted to trace the copyright holders of all material reproduced in this publication and apologize to copyright holders if permission to publish in this form has not been obtained. If any copyright material has not been acknowledged please write and let us know so we may rectify in any future reprint.

Except as permitted under U.S. Copyright Law, no part of this book may be reprinted, reproduced, transmitted, or utilized in any form by any electronic, mechanical, or other means, now known or hereafter invented, including photocopying, microfilming, and recording, or in any information storage or retrieval system, without written permission from the publishers.

For permission to photocopy or use material electronically from this work, please access www.copyright.com (http://www.copyright.com/) or contact the Copyright Clearance Center, Inc. (CCC), 222 Rosewood Drive, Danvers, MA 01923, 978-750-8400. CCC is a not-for-profit organization that provides licenses and registration for a variety of users. For organizations that have been granted a photocopy license by the CCC, a separate system of payment has been arranged.

Trademark Notice: Product or corporate names may be trademarks or registered trademarks, and are used only for identification and explanation without intent to infringe.

Library of Congress Cataloging-in-Publication Data

Names: Rathee, Sandeep, author.
Title: Friction based additive manufacturing technologies : principles for building in solid state, benefits, limitations, and applications / Sandeep Rathee, Manu Srivastava, Sachin Maheshwari, TK Kundra, Arshad Noor Siddiquee.
Description: Boca Raton, FL : CRC Press/Taylor & Francis Group, 2018. | Includes bibliographical references and index.
Identifiers: LCCN 2017057585| ISBN 9780815392361 (hardback : acid-free paper) | ISBN 9781351190879 (ebook)
Subjects: LCSH: Three-dimensional printing. | Friction stir welding.
Classification: LCC TS171.95 .R38 2018 | DDC 621.9/88--dc23
LC record available at https://lccn.loc.gov/2017057585

Visit the Taylor & Francis Web site at
http://www.taylorandfrancis.com

and the CRC Press Web site at
http://www.crcpress.com

Contents

List of Figures ix
List of Tables xiii
List of Abbreviations xv
Preface xvii
Acknowledgments xix
Authors xxi

1 General Introduction and Need of Friction Based Additive Manufacturing Techniques 1
　Foreword 1
　1.1　Introduction 2
　1.2　Need for Friction Based Additive Manufacturing Techniques 4
　1.3　Benefits of Friction Based Additive Manufacturing Techniques 5
　1.4　Content Outline 6
　References 8

2 Additive Manufacturing Technologies 11
　2.1　Introduction 11
　2.2　Historical Development and Timeline 12
　2.3　Working Principles and Additive Manufacturing Process Chain 19
　2.4　Classification of Additive Manufacturing Techniques 20
　2.5　Common Additive Manufacturing Processes 24
　2.6　Advantages and Challenges of Additive Manufacturing Processes 24
　2.7　Applications of Additive Manufacturing Technologies 30
　2.8　Metal Additive Manufacturing Techniques 33
　　　2.8.1　Limitations of Metal Additive Manufacturing 34
　　　2.8.2　Porosity 34
　　　2.8.3　Loss of Alloying Elements 35
　　　2.8.4　Cracking and Delamination 35
　2.9　Conclusion and Future Scope of Additive Manufacturing 36
　References 37

3 Friction Based Joining Techniques 41
　3.1　Introduction 41
　3.2　Historical Development of Friction Welding 41
　3.3　Friction Welding Techniques 42
　3.4　Variants of Friction Welding Techniques 43
　3.5　Hybrid Friction Based Additive Manufacturing Processes 44

v

		3.5.1	Benefits and Limitations of Friction Based Additive Techniques 54
	3.6	Conclusions.. 55	
	References 55		

4 Friction Joining-Based Additive Manufacturing Techniques.............. 59
- 4.1 Introduction 59
- 4.2 Rotary Friction Welding... 60
  - 4.2.1 Working Principles of Rotary Friction Welding................ 60
  - 4.2.2 Process Parameters Affecting Rotary Friction Welding 63
  - 4.2.3 Additive Manufacturing with Rotary Friction Welding 64
    - 4.2.3.1 Applications of Rotary Friction Welding as an Additive Manufacturing Tool 64
- 4.3 Linear Friction Welding 65
  - 4.3.1 Working Principles of Linear Friction Welding 65
  - 4.3.2 Factors Affecting Linear Friction Welding 66
  - 4.3.3 Additive Manufacturing with Linear Friction Welding 67
    - 4.3.3.1 Applications of Linear Friction Welding as an Additive Manufacturing Tool 68
- 4.4 Comparison of Rotary Friction Welding and Linear Friction Welding.. 69
- 4.5 Advantages and Limitations of Friction Welding.......................... 70
- 4.6 Conclusion... 72
- References 72

5 Friction Deposition-Based Additive Manufacturing Techniques 75
- 5.1 Introduction 75
- 5.2 Friction Deposition 75
  - 5.2.1 General Features and Experimental Results on Additive Manufacturing Using Friction Deposition 76
    - 5.2.1.1 Development of Ferrous Metal Deposits Using Friction Deposition 79
    - 5.2.1.2 Development of Nonferrous Metal Alloy Builds Using Friction Deposition........................... 80
  - 5.2.2 Benefits and Limitations of Friction Deposition 82
- 5.3 Friction Surfacing.. 83
  - 5.3.1 Working Principles of Friction Surfacing........................... 84
  - 5.3.2 Friction Surfacing Process Parameters 85
  - 5.3.3 General Features and Status of Research of Friction Surfacing–Based Additive Manufacturing Methods 89
  - 5.3.4 Benefits and Limitations 93

		5.3.5	Applications of Friction Surfacing as Additive Manufacturing Tool	93
	5.4	Conclusion		94
	References			94

6 Friction Stir Welding-Based Additive Manufacturing Techniques ... 97

	6.1	Introduction		97
	6.2	Friction Stir Welding		98
		6.2.1	Terminology Used in Friction Stir Welding	101
	6.3	Friction Stir Additive Manufacturing		102
		6.3.1	Working Principles of Friction Stir Additive Manufacturing	102
			6.3.1.1 Steps Involved in Friction Stir Additive Manufacturing	102
		6.3.2	Friction Stir Additive Manufacturing Process Variables	104
		6.3.3	General Features and Status of Research	105
			6.3.3.1 Grain Size Variation	105
		6.3.4	Defects Associated with Friction Stir Additive Manufacturing	108
	6.4	Friction Assisted Seam Welding-Based Additive Manufacturing Method		110
		6.4.1	Working Principles of Friction-Assisted Seam Welding	111
		6.4.2	Status of Research and Recent Developments	111
	6.5	Additive Friction Stir Process		114
		6.5.1	Working Principles of Additive Friction Stir	114
		6.5.2	Microstructural Characterization in Components Developed via Additive Friction Stir	114
	6.6	Machines Utilized for Friction Stir Additive Manufacturing, Friction-Assisted Seam Welding, and Additive Friction Stir		117
		6.6.1	Conventional Machine Capable of Performing Friction Stir Welding	118
		6.6.2	Customized Friction Stir Welding Machines	118
		6.6.3	Robots Designed for Friction Stir Welding	119
	6.7	Concluding Summary		119
	References			120

7 Applications and Challenges of Friction Based Additive Manufacturing Technologies ... 125

	7.1	Introduction	125
	7.2	Applications of Friction Based Additive Manufacturing Technologies	126
	7.3	Challenges of Friction Based Additive Manufacturing Technologies	132

		7.4 Conclusion	133
		References	133

8 Conclusion 135
    8.1   Introduction 135
    8.2   Concluding Summary 135
    8.3   Future Scope 138

Index 141

List of Figures

Figure 1.1	Challenges faced by AM technology	3
Figure 2.1	Concept of additive manufacturing industrial system	20
Figure 2.2	Line diagram for data flow in typical AM process.	20
Figure 2.3	Flow chart of AM.	21
Figure 2.4	Classification on the basis of bulk materials and data transfer mechanism	21
Figure 2.5	Classification of the basis of working principles.	22
Figure 2.6	Classification on the basis of fabrication methodology.	22
Figure 2.7	(a–f) Various additively manufactured implants	32
Figure 2.8	Chronological development of MAM techniques and challenges associated with them	33
Figure 2.9	Defects during MAM: (a) keyhole porosity, (b) pores owing to lack of fusion and gas-induced porosity	35
Figure 2.10	Cracks in Rene88DT superalloy fabricated via laser solid forming: (a) long crack, (b) short crack.	36
Figure 4.1	(a) Schematic of RFW variant; welding parameters as well as forces involved in (b) CDFW, (c) LFW	62
Figure 4.2	LFW process: (a) schematic, (b) four LFW phases	66
Figure 4.3	Images of LFW joints for various metals: (a) Ti-6.5Al-1.5Zr-3.5Mo-0.3Si, (b) carbon steel (medium), (c) AA 2024.	67
Figure 4.4	(a) Conventional bladed disc, (b) BLISK.	69
Figure 4.5	Proportion of materials upon which research has been reported for: (a) CDFW, (b) IFW, (c) LFW	71
Figure 5.1	Friction-deposited rough and machined surfaces of AISI 304 deposit.	76
Figure 5.2	Steps in AM of 3D components via FD.	77
Figure 5.3	Optical micrographs of: (a) as-received base metal AISI 304 steel, (b) after FD	79

Figure 5.4 (a) Macrographic image of longitudinal section of deposit produced by FD, (b) optical microstructural images of deposit, (c) SEM image taken from edge of deposit 80

Figure 5.5 Macrographic image (longitudinal section) of friction deposit of: (a) AA 5083-titanium composite. (b) AA5083-nanocrystalline CoCrFeNi composite 81

Figure 5.6 Microstructural images of composite region showing distribution of reinforcement particles: (a) optical micrograph of AA5083-titanuim, (b) SEM micrograph of AA5083-titanuim. Arrows in the figure show interfaces of different layers; (c) and (d) are optical and SEM micrographs of AA5083-CoCrFeNi composite 82

Figure 5.7 Schematic of friction surfacing..84

Figure 5.8 Steps involved in additive manufacturing of components via FS...85

Figure 5.9 Effect of rotational speed on: (a) heat input rate of coating and substrate, (b) HAZ cross-sectional area of substrate 86

Figure 5.10 Macrographic image of influence of speed of rotation on roughness and width of coating of AISI H13 over mild steel substrate 86

Figure 5.11 Effect of traverse speed on deposit variables...............................87

Figure 5.12 Effect of axial force on cross-section of coating of mild steel over same material...88

Figure 5.13 Optical micrographs of multilayer multitrack FSed deposits of mild steel; (a) good bonding at interface between layer 1-track 2, layer 1-track 3, and layer 2-track 2; (b) unbonded region between two tracks 89

Figure 5.14 (a) Scheme of layer and track arrangement in FS deposit of three layers, in the numbering system, the first number defines layer and the second number denotes the track; (b) image of multitrack (three) deposits in first layer; (c) image of multilayer FS deposit..90

Figure 5.15 Microstructural images of multitrack multilayer deposit of mild steel showing bonding between different tracks and layers: (a) optical image of layer 1-track 2 and track 3 interface and layer 1 and layer 2 interface; (b) SEM image showing excellent bonding between layer 1 and layer 2 90

List of Figures xi

Figure 5.16 Steps and images of 3D part fabrication with fully enclosed cavities..91

Figure 5.17 Optical micrographs of build fabricated by alternate deposition of alloy 316 and alloy 410 stainless steels via FS.....92

Figure 6.1 Schematic diagram of FSW...99

Figure 6.2 Joint designs for FSW: (a) simple butt joint, (b) edge butt joint, (c) T-butt, (d) lap joint, (e) multiple lap, (f) T-lap, (g) fillet joint..99

Figure 6.3 Friction stir processes based on their working and applications. 100

Figure 6.4 Schematic arrangement of FSAM. 103

Figure 6.5 Steps utilized in FSAM..103

Figure 6.6 Microhardness readings along the direction of build at two different rotational speeds; (a) 800 rpm/102 mm/min, (b) 1400 rpm/102 mm/min, (c) tensile properties of build at 1400 rpm/102 mm/min and base metal after aging, (d) work-hardening behavior of as-received and fabricated build...108

Figure 6.7 (a) Macrographic images of FSAM-fabricated build of Al 5083 alloy; (b) hardness comparison of base metal as-built components; vertical dashed line shows microhardness of base metal; (c) stress–strain curves for FSAM-fabricated built and base metal...109

Figure 6.8 Defects in a Al 7075 alloy component fabricated via FSAM...110

Figure 6.9 Diffusion phenomena showing extension of grains from one metallic layer to other in a typical AISI 304 friction seam weld...112

Figure 6.10 Schematic of FASW method 112

Figure 6.11 Friction assisted (lap) seam welds: (a) single seam weld of AA6061 (top view); (b) multilayer multitrack weld of AISI 304: (1) top view, (2) cross-sectional view, (c) cross-section.....113

Figure 6.12 Schematic arrangement of AFS...115

Figure 6.13 Microstructural images of: (a) optical microscopic (OM) images as-received IN625 filler material; (b) OM image of as-built AFS IN625, (c) sample locations along build direction of AFS IN625, (d) Euler electron back scattered diffraction (EBSD) maps showing grain size distribution along build direction of locations shown in (c), (e) number of grains and average grain sizes.................116

Figure 6.14	Conventional retrofitted vertical milling machine capable of performing FSW/FSP/FSAM/FSAW/AFS	119
Figure 7.1	Configurations that can be welded via FW	127
Figure 7.2	Typical macrographs of: (a) friction deposition, (b) friction surfacing, showing multimaterial build	129
Figure 7.3	Stiffener assembly via FSAM process for aerospace industry; (a) I-beam with transverse stiffener, (b) stringers over flattened skin panel of fuselage, (c) air foil C-S that depicts integration in fabricating ribs and stringers in wing spar web	130
Figure 7.4	Illustration of some possible combinations that can be fabricated using FSAM (a) fully gradient, (b) alternate gradient, (c) bulk composite, (d) sandwich structure, (e) tailored composite	131

List of Tables

Table 2.1	Timeline of AM Processes	14
Table 2.2	Classification on the Basis of Material Delivery System	23
Table 2.3	Classification on the Basis of ASTM F42 Committee Guidelines (AM Machines)	23
Table 2.4	Description of Individual Techniques on the Basis of ASTM F42 Committee Guidelines (Materials and Technology)	24
Table 2.5	Summary of Some Prominent AM Processes	25
Table 2.6	Comparison of Some Prominent MAM Processes	34
Table 3.1	Friction Joining and Processing Techniques	45
Table 3.2	Timeline of Friction Based Additive Manufacturing Processes	53
Table 3.3	Specific Benefits and Limitations of FATs	54
Table 4.1	Comparison of RFW and LFW	70
Table 5.1	Process Parameters and Other Details Used in Some Experimental Studies on Friction Deposition Additive Manufacturing for Different Materials	78
Table 6.1	Grain Sizes at Different Locations of WE43 Magnesium Alloy Build Fabricated Using FSAM	106
Table 6.2	Results of Tensile Test During FASW	114
Table 6.3	Types of FSW Machine with Their Features	120
Table 7.1	Current and Potential Applications of FATs	132

List of Abbreviations

3D	Three-dimensional
3DP	Three-dimensional printing
AFS	Additive friction stir
AFW	Angular friction welding
AM	Additive manufacturing
AS	Advancing side
AWS	American Welding Society
BLISK	Blade + disk
BM	Base metal
BPM	Ballistic particle manufacture
CAD	Computer aided design
CAM	Computer-aided manufacturing
CDFW	Continuous drive friction welding
CDRX	Continuous dynamic recrystallization
CNC	Computer numerical controlled
DDRX	Discontinuous dynamic recrystallization
DMD	Direct metal deposition
DMLS	Direct metal laser sintering
DRX	Dynamic recrystallization
EBM	Electron beam melting
FASW	Friction-assisted (lap) seam welding
FATs	Friction based additive manufacturing technologies
FD	Friction deposition
FDM	Fused deposition modeling
FS	Friction surfacing
FSAM	Friction stir additive manufacturing
FSATs	Friction stir additive techniques
FSBW	Friction stir butt welding
FSLW	Friction stir lap welding
FSP	Friction stir processing
FSSW	Friction stir spot welding
FSW	Friction stir welding
FW	Friction welding
FWTs	Friction-welding techniques
HAZ	Heat-affected zone
IFW	Inertia friction welding
IJP	Inkjet printing
LAM	Laser additive manufacturing
LENS	Laser-engineered net shaping
LOM	Laminated object manufacturing

LFW	Linear friction welding
MAM	Metal additive manufacturing
NZ	Nugget zone
OM	Optical microscopic
PMZ	Partially melted zone
RFW	Rotary friction welding
RPM	Rotations per minute
RPS	Rotations per second
RS	Retreating side
SGC	Solid ground curing
SLA	Stereolithography
SLM	Selective laser melting
SLS	Selective laser sintering
TL4	Technology level 4
TMAZ	Thermomechanical affected zone
TWI	The Welding Institute
UAM	Ultrasonic additive manufacturing

Preface

The exchange of ideas and effectiveness of communication are prerequisites of any learning process. In today's era of the need for high technical competence, it is the moral responsibility of every professional to share knowledge gained for societal development. This work is an attempt by the authors toward dissemination of the information and experience gathered by the authors in the domains of additive manufacturing, friction welding, and their confluence, that is, friction based additive manufacturing techniques.

The authors started their work in the field of friction based additive manufacturing techniques as part of their experimentation, but the encouraging results that were obtained inspired the authors to venture into their depths. Much lies unexplored in this domain, and the results will definitely be worth the efforts invested. Meagre literature and absence of a book in this field has inspired the authors to undertake the responsibility of coming up with the idea of the present book. This book is written in simple language so that it will be useful even for people who are venturing in the field of friction based additive manufacturing techniques for the first time.

Friction based additive manufacturing is a term coined for utilizing friction based solid-state welding/processing techniques in conjugation with advancements of layered/additive manufacturing to produce components with superior structural and mechanical properties. This is a novel manufacturing technology area for developing high structural performance components. It utilizes the principle of layer-by-layer additive manufacturing and is a major breakthrough in the domain of the solid-state metal additive manufacturing sector.

Since these techniques are basically confluence of additive manufacturing (AM) and friction welding/processing, therefore, a primary understanding of both these fields is mandatory to fully understand the concept of friction based additive techniques. To facilitate learning, this book introduces all aspects of AM necessary to outline the need for and concept of friction based AM processes. It further introduces the friction welding techniques. A timeline of friction based AM processes is then presented. Details of various friction based additive techniques, that is, linear friction welding, rotary friction welding, friction deposition, friction surfacing, friction stir additive manufacturing, friction-assisted seam welding, and additive friction stir processes are described. Applications, trends, and the future scope of these techniques are then discussed. This book is an attempt to provide an exhaustive and extensive compilation to enrich the reader's knowledge about friction based AM processes.

Rapid ongoing technical and technological advancements in the field of fabrication processes make it difficult to present an exhaustive account of various topics. The authors have, however, put their best efforts into making this book as informative and useful as possible by including each aspect, major advancement, and trend in the field of both the parent technologies and various aspects of friction based AM techniques. Enough literature has been reviewed and the practical knowledge of the authors based upon their several years of work/research experience has been utilized to write this book.

The authors sincerely hope that this book holds value for researchers, academicians, and university students who plan to pursue a research career in the field of advancements in solid-state friction based additive manufacturing. We genuinely hope that the readers can apply knowledge of the information presented to promote research and development in this field.

The authors will genuinely welcome and appreciate any queries, advice, and observations by the readers.

Sandeep Rathee
Netaji Subhas Institute of Technology

Acknowledgments

Many personal thanks to the editor, Ms. Cindy Carelli, for agreeing to publish our hard work and helping us impart a structured technical format to this manuscript. Her suggestions and advice were of immense help in enhancing the content of this manuscript and have definitely enriched the work further for the benefit of our readers. She inarguably provided one of the best professional experiences for the authors, not only in terms of her knowledge but also in terms of her efficient staff and her ability to get the work at hand completed at the earliest. We are thankful to CRC Press for taking the initiative of publishing this book belonging to a new upcoming area of the future.

The author Mr. Sandeep Rathee wishes to thank his supervisors, Prof. (Dr.) Sachin Maheshwari and Prof. (Dr.) Arshad Noor Siddiquee. Prof. Maheshwari is the epitome of simplicity, forthrightness, and strength and is a role model for him. Prof. Siddiquee is a wonderful and intelligent person who is ever ready to help his students in every possible manner. The author also wishes to thank his parents, Mr. Raj Singh Rathee and Smt. Krishna Rathee, for their unconditional support. He wants to take this opportunity to acknowledge that they have given him the opportunity to stand tall before the world despite living their entire lives in a small village and sacrificing every happiness of their own for the sake of their children. No words of appreciation can ever be enough to describe the privilege of being born to the world's best parents. The author wishes to thank his brother, Mr. Sombir Rathee, for supporting the family for the entire course of the author's indulgence in work without a single grudge and being more of a guardian than a brother. The author wishes to wholeheartedly thank his co-author and research associate, Dr. Manu Srivastava, for her constant association and help.

The author Dr. Manu Srivastava wishes to thank her director, Dr. Sraban Mukherjee (IMS Engineering College, Ghaziabad), for his inspiration and support. She also thanks her colleagues for their support. The author wishes to convey her heartfelt thanks to her supervisors, Dr. T. K. Kundra and Dr. Sachin Maheshwari. The author feels indebted and takes this platform to acknowledge that she is blessed to be under the professional guardianship of supervisors who are like parents to her. No magnitude of words can ever quantify the love and gratitude the author feels in thanking her family members, especially her late parents, Mr. Indrapal Singh Chauhan and Mrs. Saroj Chauhan; her parents-in-law, especially her father-in-law, Dr. T. K. Singh; her sisters, Mrs. Shilpi Chauhan and Mrs. Rachna Chauhan; her husband, Mr. Ramkrishna Yashaswi, who is her pillar of strength; and her child, little Yashkrishna Srivastava, who is the world's best child. Pursuing active research always sets up a pressing need for time and hard work that required her to spend much of her time, which is otherwise meant exclusively for the family and especially her child, in academic work. The author also deeply thanks her co-author and fellow researcher, Mr. Sandeep Rathee, for the constant feedback on her strengths, for his ability to teach by example, and for his valuable feedback.

The author Dr. Sachin Maheshwari wishes to acknowledge many thanks to his wife, Mrs. Monica Maheshwari; his mother; and his children for their unconditional support toward the completion of this project.

The author Dr. T. K. Kundra feels thankful to Professor P. V. M. Rao for initiating a Rapid Prototyping Centre at IIT Delhi about two decades back and who was always ready to provide the desired help and guidance in the area of additive manufacturing. Thanks to Professor P. M. Pandey also, who took over the centre after Prof. Rao and was always willing to extend his helping hand. The author is thankful to all those students who always posed a challenge to him by seeking answers to questions related to emerging technologies of additive manufacturing. He also wishes to thank his wife, Mrs. Usha Kundra, for her understanding and cooperation.

The author Dr. Arshad Noor Siddiquee wishes to thank his family and his friends for their unconditional support.

Finally, the authors devote and dedicate this work to the divine creator. They thank the Almighty for giving them the strength to bring these thoughts and understanding of concepts into physical form. The authors pray that this work may be of enough technical competence to enlighten the readers about each aspect of friction based additive manufacturing techniques.

Sandeep Rathee
Dr. Manu Srivastava
Dr. Sachin Maheshwari
Dr. T.K. Kundra
Dr. Arshad Noor Siddiquee

Authors

Sandeep Rathee is working as a teaching-cum-research fellow in the division of Manufacturing Processes and Automation Engineering at Netaji Subhas Institute of Technology, New Delhi. His PhD work is in the field of friction stir processing/welding. He has been active in the field of friction stir processing/welding and additive manufacturing for the last few years. His fields of research are friction stir welding/processing, additive manufacturing, advanced manufacturing processes, and optimization. He has over 25 publications in reputed international journals and refereed conferences. He has coauthored four book chapters in Springer books. He has total teaching experience of around seven years. He teaches the following courses at the graduate and postgraduate level: Manufacturing Processes, Welding, Casting, Workshop Technology, Advanced Manufacturing Processes, Fluid Mechanics, and so on. He is a lifetime member of the Additive Manufacturing Society of India (AMSI), and Vijnana Bharati (VIBHA).

Dr. Manu Srivastava is currently serving as an associate professor in the Department of Mechanical Engineering, IMS Engineering College, Ghaziabad. She has completed her PhD in the field of additive manufacturing. She has been active in the field of additive manufacturing research for the last eight years. Her areas of research include additive manufacturing, friction stir processing, friction based additive manufacturing, automation, manufacturing practices, and optimization techniques. She has over 35 publications in international journals of repute and refereed international conferences. She has four chapters in Springer, Verlag series books. She has a total of 14 years of experience in teaching and research. She has won several proficiency awards during the course of her career, including merit awards, best teacher awards, and so on. She teaches the following courses at the graduate level: Manufacturing Technology, Advanced Manufacturing Processes, Material Science, CAM, Operations Research, Optimization Techniques, Engineering Mechanics, Computer Graphics, and so on. She is a lifetime member of the Additive Manufacturing Society of India, Vijnana Bharati (VIBHA), The Institution of Engineers (IEI India), the Indian Society for Technical Education (ISTE), and Indian Society of Theoretical and Applied Mechanics (ISTAM).

Dr. Sachin Maheshwari is currently serving Netaji Subhas Institute of Technology as a senior Professor and Head of the Department in the division of Manufacturing Processes and Automation Engineering. Additionally, he is also the nominated Dean, Faculty of Technology (Delhi University) and IRD (NSIT). He is a known name in the field of advanced welding and manufacturing processes. He has completed his PhD from the Indian Institute of Technology, Delhi, in the field of welding and holds an ME in industrial metallurgy from the Indian Institute of Technology, Roorkee. His areas of interest include all variants of welding; advanced manufacturing, computer graphics, additive manufacturing; optimization techniques; and unconventional manufacturing processes. He has over 80 research papers in international journals and refereed conferences. He has guided several PhD theses. About 10 PhD theses have been awarded and equal numbers are in progress. He has two patents. He has a total teaching and research experience of around 23 years and has taught a wide assortment of subjects during his teaching career. He is a seasoned academician who has made his mark in mechanical engineering during the last decade. He has rich experience of working on statutory authorities and experience with handling academic assessment and accreditation procedures. He is associated with many research, academic, and professional societies in various capacities.

Dr. T. K. Kundra has served the Indian Institute of Technology Delhi as professor and head of the department of Mechanical Engineering and continues to serve the same world-renowned institution. His areas of interest are optimal mechanical system design, including microsystems and computer-integrated manufacturing systems including additive manufacturing. His experience of teaching, research, and design is spread over 48 years and includes teaching at AIT Bangkok and Addis Ababa University, and studies at Loughborough University, Imperial College, Ohio State University, and TU Darmstadt. He has been consulted by several organizations such as Hero Motors, BHEL, Eicher, ONGC, DRDO, the British Council, and so on. He is a chartered engineer and Fellow of the Institution of Engineers. He is author/coauthor of about 100 technical papers, coauthor of two textbooks on numerical control and computer aided manufacturing, and author of a book on optimum dynamic design. He has also been awarded the honor of Mechanical Engineer of Eminence. He has introduced/developed/taught a wide spectrum of subjects (around 40) in his teaching career at the graduate/postgraduate level, including mechanical design, optimization, plant equipment design including computer numerical controlled (CNC) manufacturing and so on. He is associated with many professional societies in different capacities.

Arshad Noor Siddiquee works as a Professor, Department of Mechanical Engineering, Jamia Millia Islamia (A Central University), Delhi, India. He received his PhD and MTech from the Indian Institute of Technology, Delhi, India, after graduating from Government Engineering College, Jabalpur, India. His research areas include welding, conventional and nonconventional manufacturing processes, advance manufacturing, friction stir welding/processing, etc. He has taught a wide spectrum of subjects in his teaching career of more than 20 years. He has authored/coauthored over 100 research papers in reputed international journals. He has authored/coauthored five books, four monographs, and three patents. He is guiding several PhD scholars. He is leading a funded research project worth several million Indian rupees on friction stir welding/processing. He is associated with many professional societies in various capacities.

1
General Introduction and Need of Friction Based Additive Manufacturing Techniques

Foreword

People tend to doubt the effectiveness and working of friction based additive manufacturing technologies (FATs) when they hear about them for the first time. The usual reactions that are encountered include "Do these processes work?" "How do they work?" "What are the prerequisites of FATs?" "Do these processes involve complex procedures?" The present book is a sincere endeavor by the authors to answer these research queries. Actually, these processes are very simple in operation and produce impressive results in terms of improved microstructures as well as enhanced mechanical and structural properties. Observing the working of these processes is quite intriguing, where there is a rotational component in the form of a consumable/nonconsumable rod/part and a consumable/nonconsumable metallic tool. The joining of two metallic rods/surfaces, the deposition of consumable rod over substrate, the joining of parts via a nonconsumable rotating tool without fusion, and the deposition of material through a hollow tool have the ability to capture technical inquisitiveness and attract anyone. The realization of additive manufacturing of three-dimensional (3D) parts via these techniques is also quite interesting. During these processes, there are no fumes, melting, or emanation of affluent lines, hazardous gases, or radiation. Noise during these processes is also low. This leads people to normally disbelieve or underestimate the capabilities of these innovative techniques and makes it hard to assimilate the fact that these processes can fabricate 3D components easily in shorter cycle times and with enhanced properties as compared to base metals. In general, these techniques utilize principles of friction based processes to build a 3D part through layer by layer additive manufacturing. Some of these processes are old, however, their AM versions are quite recently developed.

The primary objective of this book is to articulate the process principles of different FATs. This includes description of process parameters involved and their effect upon process efficiency, fabricated part characteristics, general features, and process-specific limitations, done in a simple, logical, and concise way. As these processes are new and evolving, based on current knowledge of these techniques and available literature, applications of these techniques in different sectors are described. Toward the end, a brief conclusion of these techniques is presented, followed by discussion on their future prospects.

1.1 Introduction

The consistent need to reduce weight in the aerospace, marine, and automobile sectors has always been an area of key interest among researchers. In recent times, utilizing lightweight materials has been found to enhance the performance of components in such applications. For example, there is an approximate annual reduction of hundreds of gallons of fuel usage in air carriers and carbon dioxide emission using lightweight materials [1,2]. In today's technical scenarios, optimal fabrication of such lightweight components with improved properties using a suitable process is a key focus of research. Moreover, the ease with which complex shaped components can be fabricated is useful in improving the effectiveness and efficiency of any process used to produce high-end components for specific applications. One of the most suitable techniques for fulfilling such a need is additive manufacturing.

AM is an advanced and highly established three-decade-old technology for fabrication of various complex shaped parts with little effort as compared to conventional manufacturing techniques. Apart from being used for the 3Fs, that is, form, fit, and functional applications, AM techniques are applied to process chains for minimizing cost and time requirements. AM refers to a class of manufacturing where three-dimensional parts are fabricated via 2.5 dimensional layer additions directly from CAD designs by utilizing different strategies. These techniques require minimal human intervention and are appreciably economical. In addition to these benefits, several comprehensive studies have also established that wastage in using AM techniques is appreciably less and they are more environmentally friendly as compared to conventional processes.

Despite the manifold advantages of AM techniques, they suffer from some inherent technical limitations and challenges, as illustrated in Figure 1.1. Lots of research is in progress to address these challenges.

AM is currently used for both metallic and nonmetallic raw materials. A huge quantum of research has been accomplished on nonmetallic materials as compared to metals. Metal-based additive manufacturing (MAM) is still in its development stage. MAM has been developed for only a few metals, and poor lateral strength is a serious issue even in them [4,5]. Parts manufactured using MAM possess the inherent disadvantage of anisotropy

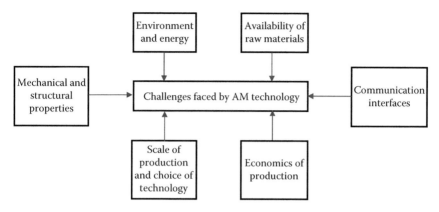

FIGURE 1.1
Challenges faced by AM technology. (Adapted from Srivastava, M. et al. Design and processing of functionally graded material: Review and current status of research, in *7th International Conference on 3D Printing & Additive Manufacturing Technologies- AM 2017, Global Summit* 2017, Bengaluru, India.) [3]

and low transverse strength, chiefly owing to accompanying liquidus-solidus phase transformations. This renders the parts unsuitable for structural applications, thereby putting restrictions on utilization of structural parts manufactured using MAM [6,7]. Popular MAM techniques like electron beam melting, selective laser melting, and so on are highly cost intensive, too. In order to fulfill demands from various sectors like aerospace, automotive, and tooling, the focus of AM research has shifted toward the development of processes/methods for AM of metallic components. Any improvements in MAM technology to troubleshoot the inherent fabrication problems mentioned in the foregoing text will have a huge significance in improving the usage of such a game-changing process. Thus, there is a strong need for a process that can utilize the fast development time and ease of fabrication of AM and also simultaneously address the anisotropy-related issues of the metallic components. In line with concerns raised above, numerous studies have been carried out to alleviate these limitations. One feasible solution is the incorporation of solid-state friction based approaches into AM. These processes are hybrid in nature and work on the layer-by-layer principle along with friction based joining. There are different innovative approaches to friction based additive techniques (FATs) and can basically be categorized into seven types: rotary friction welding (RFW), linear friction welding (LFW), friction deposition (FD), friction surfacing (FS), friction stir additive manufacturing (FSAM), friction assisted (lap) seam welding (FASW), and additive friction stir (AFS) [8–15]. All of these techniques are basically variants of friction welding. The friction joining of materials in these processes may take place directly or indirectly. In RFW and LFW, direct friction welding takes place wherein the addition of material occurs in the form of joining of two surfaces (rod forms in RFW and rectangular or other shapes in LFW).

For the realization of 3D parts, joined parts are machined into sliced contours using CNC machining, and these steps are repeated until a desired build height is achieved. In FD and FS, deposition of material takes place from a consumable rod rotating against the substrate. In FSAM, FASW, and AFS, addition of material is accomplished following the principles of the friction stir welding (FSW) process, which is also a variant of friction welding.

1.2 Need for Friction Based Additive Manufacturing Techniques

The ability of MAM techniques to be used for the fabrication of intricate parts has led to AM being considered an option for replacing parts fabricated via conventional manufacturing, especially in the aviation, prosthesis, biomedical, automotive, and marine sectors. Several MAM techniques are in practice that generally utilize powder and a metal wire or sheet as feedstock material for their consolidation into dense metallic components by application of suitable energy sources like electron beam, laser, electric arc, ultrasonic vibration, and so on [16]. These techniques can be categorized into four main heads (as per ASTM standard terminology for AM techniques [17]): processes based on powder bed fusion like electron beam melting, selective laser sintering, and so on; processes based on directed energy deposition like laser consolidation, arc additive manufacturing, and so on; processes based on binder jetting such as inkjet 3D printing; and processes based on sheet lamination such as ultrasonic consolidation, laminated object manufacturing, and so on [18]. Processes based on the first three categories utilize either metal powder or metal wire as feedstock material, and most of these processes generally rely on liquid phase processing. Owing to this, there are several issues related to fusion-based processing that limit its utility in certain applications. These include the fact that during fusion-based processes, the powder particle-based substrate and particles are in an activated state due to high surface energy content, making them vulnerable to contamination. This can be further explained as follows: during AM, the substrate is in a molten state, and it is more prone to form discontinuities such as internal porosity, inclusions, and other solidification defects. Also, metal wire–based processes are prone to fabricating inhomogeneous microstructures. These processes can be considered equivalent to microcasting or microwelding. The processes based on fourth category, that is, sheet lamination processes, utilize metallic foil/sheet/laminate as feedstock material and are still in the nascent stage [16]. However, ultrasonic additive manufacturing (UAM) reduces the defects occurring owing to solid liquid phase transformations owing to its solid-state nature [19,20]. It has the ability to fabricate multimaterial components in the solid state and is largely utilized for AM of metallic materials. However, it suffers from its own limitations of inhomogeneous microstructures at interfacial and noninterfacial regions, resulting in inferior properties as compared to base material [21].

Thus, the issues in the fabrication of metallic components using fusion-based AM techniques can be summarized as: poor structural strength, porosity, shrinkage defects, inclusions, and other solidification defects due to liquid-solid phase transformations, powder contamination, inhomogeneity in microstructures and directionally variant mechanical properties, poor properties along the build direction owing to processing misalignments, the stair-stepping effect, restricted production volume, requirement of closely controlled chambers, limitations on the number of possible alloys, and so on. For successful utilization of MAM in biomedical, transportation, aviation, and other critical sectors, it is mandatory to overcome the aforementioned challenges, which will then open altogether new avenues for MAM. This is in turn related to devising some methods to obtain superior microstructures and discontinuity-free parts fabricated via MAM. These challenges necessitate the requirement to focus ongoing research upon issues beyond intricate shapes and try to develop components that can meet the strict strength restrictions of user industry requirements.

This demands incorporation of solid-state friction based approaches into AM, which brings FATs into the main role. These processes utilize the layer-by-layer building principle of AM with the strength enhancement ability of friction joining, thereby resulting in a rare excellent choice and flexibility to get complicated robust components in a limited time. Thus, FATs can be considered as a special innovative class of novel, fast, and size-independent enabling MAM techniques. They have been successfully employed for fabrication of simple shaped objects by various researchers. They can be utilized to produce fully dense and near net shape wrought graded components economically from a broad range of alloys with graded structures.

1.3 Benefits of Friction Based Additive Manufacturing Techniques

The specific advantages of FATs as compared to fusion-based MAM are their considerably low-energy consumption, optimal part consolidation, and structural efficacy. It has been established by various research studies that friction based AM processes consume only around 2.5% of the energy consumed by fusion-based processes [22]. A perspective representation of fusion-based MAM and FATs was proposed by Palanivel et al. [8], and states that FATs consume less energy (power requirement) and produce less distorted parts with high structural performance as compared to fusion-based AM processes. In addition, the appreciably high heat flux required in fusion-based AM processes adds to woes such as porous structures, shrinkage cavities, and so on. All of the aforementioned FATs are free from/less prone solidification defects. Fewer distortional defects, lower porosities, the ability to fabricate larger components, multimaterial bonding abilities, greater reproducibility rates, excellent metallic properties, tailor-made microstructures, and so on are major benefits of friction based AM techniques as compared to fusion-based MAM processes.

Thus, depending on their solid-state nature, these techniques have several advantages, which can be summarized as follows:

- Green manufacturing technologies
- Eliminate the majority of fusion-based defects
- Utilize less energy as compared to fusion-based MAM techniques
- Can be used for a broad material range of ferrous and nonferrous metals
- Produce fine-grained equiaxed microstructures
- Produce high-strength components as compared to base materials
- Engineered microstructures possible
- Dissimilar material builds can be easily manufactured using these techniques
- Limitations on the size of components that can be fabricated via conventional AM processes can be easily overcome because of the absence of build volume restrictions.

1.4 Content Outline

This book provides comprehensive details on newly developed hybrid solid-state FATs. The main aim of the book is to enhance the spectrum of reader interest toward understanding the concept of these intriguing hybrid processes apart from AM and core friction based techniques fields. Also, necessary information about parent processes (AM and FW) is provided in a single platform to fully understand the basics of FATs. The authors have attempted to provide comprehensive details of these processes in terms of process principles, present research status, advancements, industrial applications, and future scope of these techniques. However, these details are not exhaustive owing to the restriction on scope of the present work, and readers with a keen interest are advised to refer to the references for a particular process for more details.

The book consists of eight chapters and is organized as follows:

Chapter 1, *Introduction*, presents the various aspects of AM, friction welding, and their confluence, that is, FATs. This chapter presents an overview of various aspects/chapters addressed in this book.

Chapter 2, *Additive Manufacturing (AM) Technologies*, briefly describes the major commercial AM processes, their timelines, and recent developments. The issues and challenges in fusion-based MAM techniques are systematically described.

Chapter 3, *Friction Based Joining Techniques*, introduces various friction based processes and describes the potential techniques that can be effectively utilized for AM of metallic materials. A timeline of these techniques is presented for clarity of evolution of these processes.

Chapter 4, *Friction Joining-Based Additive Manufacturing Techniques*, defines two friction welding-based AM processes (i.e., rotary friction welding and linear friction welding), which can be successfully used for AM of metallic parts. The current status of research, their working principles, benefits, and limitations of these techniques are discussed. A few interesting applications of these techniques are also discussed.

Chapter 5, *Friction Deposition-Based Additive Manufacturing Techniques*, addresses deposition-based friction AM techniques. These are mainly termed FD- and FS-based AM techniques. The working principles, effect of process parameters, benefits, limitations, and current status of research of both processes are discussed. Various intriguing applications of these techniques are also illustrated.

Chapter 6, *Friction Stir Welding-Based Additive Manufacturing Techniques*, addresses friction stir welding-based AM techniques, that is, FSAM, FASW, and AFS. An introduction to friction stir welding; terminology used in this process; and evolution of microstructure during FSAM, FASW, and AFS are described. Defect formation during these processes and their causes are explained, with suitable illustrations. Machines utilized for these techniques and the effect of process parameters are discussed in detail.

Chapter 7, *Applications and Challenges of Friction Based Additive Manufacturing Technologies*, presents a detailed discussion of the present as well as the proposed applications of FATs. The challenges to accomplish full-scale application of these innovative techniques are discussed in a concise way.

Chapter 8, *Conclusion*, concludes the book based upon analysis accomplished in the previous chapters. Future scope/trends are also highlighted toward the end of this chapter.

References

1. Hawkins, S. Fuel smart, in *Ascend* 2016, https://www.sabreairlinesolutions.com/images/uploads/FuelSmart.pdf
2. Phinazee, S. Efficiencies: Saving time and money with electron beam free form fabrication. *Fabricator*, 2007. 15:20.
3. Srivastava, M., Rathee, S., Maheshwari, S., Kundra, T. K. Design and processing of functionally graded material: Review and current status of research, in *7th International Conference on 3D Printing & Additive Manufacturing Technologies- AM 2017, GLOBAL SUMMIT* 2017, Bengaluru, India.
4. Frazier, W. E. Metal additive manufacturing: A review. *Journal of Materials Engineering and Performance*, 2014. 23(6): 1917–1928.
5. Ahn, D.-G. Direct metal additive manufacturing processes and their sustainable applications for green technology: A review. *International Journal of Precision Engineering and Manufacturing-Green Technology*, 2016. 3(4): 381–395.
6. Safronov, V. A., Khmyrov, R. S., Kotoban, D. V., Gusarov, A. V. Distortions and residual stresses at layer-by-layer additive manufacturing by fusion. *Journal of Manufacturing Science and Engineering*, 2016. 139(3): 031017–031017-6.
7. Sames, W. J., List, F. A., Pannala, S., Dehoff, R. R., Babu, S. S. The metallurgy and processing science of metal additive manufacturing. *International Materials Reviews*, 2016. 61(5): 315–360.
8. Palanivel, S., Mishra, R. S. Building without melting: A short review of friction-based additive manufacturing techniques. *International Journal of Additive and Subtractive Materials Manufacturing*, 2017. 1(1): 82–103.
9. Dilip, J. J. S., Janaki Ram, G. D., Stucker, B. E. Additive manufacturing with friction welding and friction deposition processes. *International Journal of Rapid Manufacturing*, 2012. 3(1): 56–69.
10. Dilip, J. J. S., Babu, S., Varadha Rajan, S., Rafi, K. H., Janaki Ram, G. D., Stucker, B. E. Use of friction surfacing for additive manufacturing. *Materials and Manufacturing Processes*, 2013. 28(2): 189–194.
11. Dilip, J. J. S., Kalid, R. H., Janaki Ram, G. D. A new additive manufacturing process based on friction deposition. *Transactions of the Indian Institute of Metals*, 2011. 64(1): 27.
12. Kandasamy, K. Solid state joining using additive friction stir processing. 2016. Google Patents.
13. Palanivel, S., Sidhar, H., Mishra, R. S. Friction stir additive manufacturing: Route to high structural performance. *JOM*, 2015. 67(3): 616–621.
14. Rodelas, J., Lippold, J. Characterization of engineered nickel-base alloy surface layers produced by additive friction stir processing. *Metallography, Microstructure, and Analysis*, 2013. 2(1): 1–12.
15. Kalvala, P. R., Akram, J., Tshibind, A. I., Jurovitzki, A. L., Misra, M., Sarma, B. Friction spot welding and friction seam welding. 2014. Google Patents.
16. DebRoy, T., Wei, H. L., Zuback, J. S., Mukherjee, T., Elmer, J. W., Milewski, J. O., Beese, A. M., Wilson-Heid, A., De, A., Zhang, W. Additive manufacturing of metallic components—Process, structure and properties. *Progress in Materials Science*, 2018. 92(Supplement C): 112–224.

17. ASTM-Standard. Standard terminology for additive manufacturing technologies, F2792-12a. 2013.
18. Ding, D., Pan, Z., Cuiuri, D., Li, H. Wire-feed additive manufacturing of metal components: Technologies, developments and future interests. *The International Journal of Advanced Manufacturing Technology*, 2015. 81(1): 465–481.
19. Janaki Ram, G. D., Robinson, C., Yang, Y., Stucker, B. E. Use of ultrasonic consolidation for fabrication of multi-material structures. *Rapid Prototyping Journal*, 2007. 13(4): 226–235.
20. Obielodan, J. O., Ceylan, A., Murr, L. E., Stucker, B. E. Multi-material bonding in ultrasonic consolidation. *Rapid Prototyping Journal*, 2010. 16(3): 180–188.
21. Dehoff, R. R., Babu, S. S. Characterization of interfacial microstructures in 3003 aluminum alloy blocks fabricated by ultrasonic additive manufacturing. *Acta Materialia*, 2010. 58(13): 4305–4315.
22. Baumann, J. A. *Technical Report On: Production of Energy Efficient Preform Structures* 2012, The Boeing Company: Huntington Beach, CA.

2
Additive Manufacturing Technologies

2.1 Introduction

Today, global markets are competitive, cutting edge, and customer oriented, and face an alarmingly accelerated growth rate of obsolescence. This requires manufacturing firms to customize products and save costs to achieve consistent business upgrades in order to stay progressive [1]. AM is a three-decade-old and still-evolving technology that has proven versatility in design visualization and manufacturing advancement [2]. It has been a manufacturing methodology of huge significance over the past few decades. A number of research studies have shown that AM techniques have the ability to elicit the next industrial revolution [3,4]. AM can be understood as the process of joining materials in layers for component fabrication from 3D models/component data. As per ASTM standards, AM is a method to join raw materials to obtain components as compared to the conventional method of component fabrication by subtracting material from bulk. This definition is applicable to all AM systems including metallic, ceramic, polymer, and so on. [5]. Though the general trend in AM is addition of material in a layered fashion, sometimes there can be variations like subtractive additive manufacturing or adding material in voxels rather than in layers (as in ballistic particle manufacture). Different physical and chemical phenomena by which layers are subsequently added in each AM process have been comprehensively reviewed by various researchers [6–8].

AM is known by different key names like additive (layer/digital) manufacturing, layered (based/oriented) manufacturing, rapid (prototyping, tooling, and manufacturing) technology, digital (fabrication/mock-up) technology, direct (tooling and manufacturing) technology, 3D (printing/modeling) technology, desktop, solid freeform, generative, on-demand manufacturing, and so on [9]. It forms a critical pillar of worldwide manufacturing together with conventional manufacturing techniques (both subtractive and formative). To understand the advent and development of various AM processes, a brief discussion of their timeline is presented in the next subsection.

2.2 Historical Development and Timeline [10–14]

This section aims to present a picture of the timeline and gradual development of AM technology. The first patent in this field was filed by Dr. Kodama in Japan in May 1980. However, the visible origins of AM were in 1986, when Dr. Charles Hull invented a stereolithography (SLA) machine. In 1987, Carl Deckard, University of Texas, filed a US patent for selective laser sintering (SLS), which was granted in 1989. Later, a license was acquired by 3D Systems. In 1989, Dr. Scott Crump filed a patent for fused deposition modeling (FDM), and the patent was issued to Stratasys in 1992. This technology is used by many entry-level machines based on the self replicating or RepRap principle. In 1989, EOS GmbH was founded in Germany by Hans Langer, and its focus was on laser sintering, which is known for quality prototyping and functional and production applications. Its stereos and direct metal laser sintering (DMLS) systems were considerably efficient. Other Rapid Prototyping (RP) systems that emerged around the 1990s were ballistic particle manufacture (BPM) by William Masters, laminated object manufacturing (LOM) by Michael Feygin/Helisys, solid ground curing (SGC) by Ifz Chak Pomerant et al./Cubital, and three-dimensional printing (3DP) by Emaneul Sachs et al./3D Systems. Out of all these systems, only 3D Systems, Stratasys, and EOS have survived.

Later, a shift was witnessed in AM systems from research, development, and prototyping toward more industrial and functional components and terms like rapid tooling and manufacturing, which are in wide usage today. Technologies like Sanders Prototype in 1996, Z Corporation in 1996, Arcam in 1997, Objet Geometries in 1998, SLM in 2000, Envision Tec in 2002, and EBW/Exone in 2005 emerged over time. During the mid-1990 s, the AM sector witnessed a demarcation along two discrete directions. The first included high-end 3D modelers, which were used for fabrication of expensive, highly engineered and complex, and high-value products. This enabled using AM for aerospace, automotive, medical, and fine jewelry applications. The second direction was concept modelers, which mainly emerged owing to raging price competitions. These were mainly for functional prototypes and developing concepts, and were office friendly and cost effective.

Gradually, owing to improvement in prices, speed, material choices, and accuracy, 3D Systems was a pioneer in developing its first system under $10,000. This was followed by the conceptualization of a $5000 desktop factory that did not materialize. One landmark in the field of AM was the concept of self-replicating machines, or RepRap technology, by Dr. Bowyer, which revolutionized the entire AM market. This amounted to open-source technology becoming an industrial practice where everyone who cared had access to and an idea about 3D printers. In 2009, the first RepRap 3D printer became available on the market. The second RepRap printer was commercialized in 2009 by MarketBot. Since then, these printers are amply available on the market. In 2012, alternative 3D processes came into being via B9 creations. Also, the Form 1 modeler materialized during the same time. This development is still continuing, and newer advanced technologies and modelers are booming, amounting to market diversity in terms of technological compatibility and competence. A detailed timeline of these AM technologies and modelers is presented in Table 2.1.

The growth of AM technologies has witnessed tremendous progress and followed specific trends over the last few years. Campbell et al. [15] made a few important predictions for the AM trends over the next few years, which were:

1. Expiry of primary patents would result in many low-/medium-cost AM modelers, which will be easily available to the public, leading to more demand for these systems. Also, newer improved raw materials will lead to increased supply and demand.
2. Fabricating speed will demonstrate significant improvement with the material and design advancements, thereby resulting in availability of physical parts in minutes rather than hours.
3. Future AM machines based on hybrid AM will result but will have reduced versatility.
4. Future AM systems (not based on all existing technologies) shall possess multimaterial processing capabilities.
5. AM systems have immense potential in tissue engineering.

All these predictions are more or less true in the present context of developments in AM. It should be kept in mind that AM is not supposed to completely replace conventional manufacturing practices. However, it will have unique benefits in various domains. Huge growth in the AM sector and in turn appreciable savings in manufacturing as well as material costs have been predicted over the next decade by various researchers. This is clearly visible from the growth trends in the last few years. The growth has mainly been systems and application based rather than technology or modeler based. Eyers et al. [16] have suggested a newer concept of industrial additive manufacturing systems (IAMS), whose general configuration would be as illustrated in Figure 2.1 and can be defined as possessing enough maturity to produce prototypes, tooling systems, components, or complete products in actual world manufacturing domains.

The expiry of the FDM patent has, in itself, been a landmark. The most important change in general thinking since AM is now being seen in the consumption habits and manufacturing imaginations of customers, which is evident from various manufacturing and policy reports [15–17]. Numerous researchers, institutions, centers, and industries have now fully incorporated AM into their innovation, research, and manufacturing habits. A completely additively manufactured car in 2010, a food printing modeler in 2011, the first off-Earth space printing machine purchased by NASA in 2014, carbon clip techniques in 2015, concrete printing and bone printing in 2017, and so on reflect the degree of innovative trends and applications of AM over recent years [17–19].

TABLE 2.1

Timeline of AM Processes

Sr. No.	Year of Advent	Name of the Process/Modeler	Patented/Commercialized By
1	1987	SLA	3D Systems
2	1991	FDM	Stratasys
		Solid ground curing (SGC)	Cubital
		LOM	Helisys
3	1992	SLS	DTM, which is currently a unit of 3D Systems
		Solidform stereolithography system	DuPont/Teijin Seiki
4	1993	Direct shell production casting	Soligen
		Denken's SL system	Denken
		QuickCast	3D Systems
5	1994	ModelMaker	Solidscape/Sanders Prototype
		Meiko's stereolithography machine	Meiko
		Solid Center—LOM	Kira Corp
		Fockele & Schwarze stereolithography machine	Fockele & Schwarze
		EOSINT	EOS
6	1995	Ushio stereolithography machine	Ushio
7	1996	Genisys machine	Stratasys
		3D printer—Actua 2100	3D Systems
		3D printer—Z402	Z Corp.
		Semiautomated paper lamination system	Schroff Development
		Personal Modeler 2100—BPM	BPM Technology
		DuPont's Somos stereolithography technology	Aaroflex
		Zippy paper lamination systems	Kinergy
8	1997	Laser additive manufacturing	AeroMet
9	1998	Beijing Yinhua laser rapid prototype making & mold technology	Tsinghua University
		E-DARTS stereolithography system	Autostrade
		Laser-engineered net shaping	Optomec
10	1999	ThermoJet	3D Systems
		ProMetal RTS-300 machine	Extrude Hone AM business
		Steel powder-based selective laser melting system	Fockele & Schwarze
		Controlled metal buildup machine	Röders
11	2000	Rapid Tool Maker	Sanders Design International
		Color 3D printer	Buss Modeling Technology
		Quadra	Objet Geometries
		PatternMaster	Sanders Prototype—Solidscape
		Direct metal deposition	Precision Optical Manufacturing

(Continued)

TABLE 2.1 (*Continued*)
Timeline of AM Processes

Sr. No.	Year of Advent	Name of the Process/Modeler	Patented/Commercialized By
12	2001	Ultrasonic consolidation process	Solidica
		Perfactory machine based on digital light processing	Envisiontec
		EuroMold 2001	Concept Laser GmbH
13	2002	Direct metal deposition (DMD)	POM group
14	2003	Z Printer310	Z Corp
		Sony stereolithography machines	Sony Precision Technology
		T612 system	Solidscape
		InVision 3D printer	3D Systems
		Low-cost Wizaray stereolithography system	Chubunippon
		EOSINT M 270	EOS group
		TrumaForm LF and DMD505	Trumpf
15	2004	3 versions of FDM Vantage	Stratasys
		Vanquish photopolymer-based System	Envisiontec
		RX-1 metal-based machine	ProMetal division of Ex One
		InVision HR, Sinterstation HiQ	3D Systems
		Ultrasonic consolidation system—Formation machine	Solidica
		Dual-vat Viper HA stereolithography system	3D Systems
		Vero FullCure 800 series	Objet Geometries
		EOSINT P 385	EOS
		M1 cusing laser melting machine	Concept Laser
		DigitalWax 010 and 020 systems	Next Factory (now DWS)
		T66 and T612 Benchtop systems	Solidscape
16	2005	Spectrum Z510	Z Corp.
		Sinterstation Pro	3D Systems
		InVision LD	Solido and rebranded by 3D Systems
		SEMplice laser-sintering machine	Aspect Inc.
		Eden500V	Objet Geometries
		ZPrinter 310 Plus	Eden500V
		Large Viper Pro SLA	3D Systems
		Sand-based Sprint machine	Ex One's ProMetal division
		SLM ReaLizer 100	MCP Tooling Technologies
		VX800 machine	Voxeljet Technology GmbH
17	2006	Eden350/350V platform, Eden250 3D	Objet Geometries
		InVision DP	3D Systems
		Vantage X systems	Stratasys

(*Continued*)

TABLE 2.1 (*Continued*)

Timeline of AM Processes

Sr. No.	Year of Advent	Name of the Process/Modeler	Patented/Commercialized By
		Dimension 1200 BST and SST	Stratasys
		NanoTool	DSM Somos
		Z Scanner 700 handheld 3D scanner	Z Corp.
		Formiga P 100 laser-sintering system, EOSINT P 390 and EOSINT P 730	EOS group
		VX800 machine	Voxeljet Technology
		Small Perfactory Desktop System	Envisiontec
		SLM ReaLizer 100 selective laser-melting machine	ReaLizer (Germany based)
18	2007	V-Flash 3D printer	3D Systems
		Dimension Elite 3D printer	Stratasys
		D66 and R66 T66 machines	Solidscape
		ZPrinter 450	Z Corp.
		A2 Electron beam melting	Arcam
		Formiga P 100 laser-sintering system	EOS
		FDM 200mc machine	Stratasys
		M2 cusing system	Concept Laser
		FDM 400mc	Stratasys
		VX500 system	Voxeljet
		FDM 900mc	Stratasys
		PerfactoryXede	Envisiontec
		Connex500 3DP system	Objet Geometries
		FigurePrint approach	Microsoft
		Pollux 32 selective mask-sintering system	Sintermask
19	2008	FDM 360mc	Stratasys
		ProJet HD3000	3D Systems
		Dimension 1200es 3D printer	Stratasys
		V-Flash desktop modeler	3D Systems
		T76 precision wax printing System	Solidscape
		LENS MR-7 machine	Optomec
		iPro 9000 SLA Center stereolithography system	3D Systems
		ProJet SD 3000 3D printer	3D Systems
		PolyStrata microfabrication technology	Nuvotronics
		TetraLattice Technology	Milwaukee School of Engineering
		iPro 9000 XL SLA, large-format ProJet 5000, sPro 140 and 230 SLS Centers, iPro 8000 MP SLA Center, ProJet CP 300 RealWax 3D printer	3D Systems

(*Continued*)

Additive Manufacturing Technologies 17

TABLE 2.1 (*Continued*)

Timeline of AM Processes

Sr. No.	Year of Advent	Name of the Process/Modeler	Patented/Commercialized By
		SLM 250-300	MTT
		Matrix 3D printer	Mcor Technologies
		Micro Light Switch technology	Huntsman Advanced Materials
		ZPrinter 650	Z Corp
		Alaris30 PolyJet machine, Eden260V machine	Objet Geometries
		Araldite Digitalis	Huntsman Advanced Materials
		EOSINT P 800	EOS
		Fortus Finishing Stations	Stratasys
		Spore Sculptor facility	Electronic Arts, Z Corp.
		JuJups.com	Genometri of Singapore
		Eden260V machine	Objet Geometries
20	2009	ASTM Committee F42 on AM technologies established	ASTM International headquarters
		uPrint Personal Printer	Stratasys
		Shapeways Shops	Shapeways
		SLM 50	ReaLizer GmbH
		RapMan 3D printer Kit	Bits from Bytes
		Cupcake CNC product	MakerBot Industries
		ULTRA bench-top DLP-based system	Envisiontec
		PreXacto line	Solidscape
		VFlash 3D printer	3D Systems
		Connex350 system	Objet Geometries
		Dimatix Materials Printer DMP-3000	Fujifilm Dimatix
		Standard terminology on AM technologies	ASTM International Committee F42
		Monochrome ZPrinter 350 machine	Z Corp
		EOSINT P 395 and EOSINT P 760	EOS
		ProJet 5000	3D Systems
		SD300 Pro system	Solido
		Visible light photopolymer-based 3D printers	Carima
		VX800HP	Voxeljet
		Solidscape users group	Steven Adler of A3DM
		Design-your-own toy website launched	Keith Cottingham
21	2010	uPrint Plus	Stratasys
		Aerosol Jet Display Lab System	Optomec
		Cometruejet	Microjet Technology
		EasyCLAD Systems for its laser metal deposition equipment	Irepa Laser

(*Continued*)

TABLE 2.1 (*Continued*)

Timeline of AM Processes

Sr. No.	Year of Advent	Name of the Process/Modeler	Patented/Commercialized By
		Dental manufacturing facility	Renishaw plc (UK)
		Consumer-oriented products made from glass	Shapeways
		HP-branded FDM machines entered into market	Stratasys
		ZBuilder Ultra	Z Corp.
		Portable personal UP! 3D printer	Delta Micro Factory Corp.
		Monochrome ZPrinter 150 and color ZPrinter 250	Z Corp.
		Wax drop-on-demand printers	Solidscape
		DMLS in rapid tooling	EOS and GF AgieCharmilles
		Service named InvisHands started	INUS
		EOSINT M 280 system	EOS
		Objet24 3D printer	Objet Geometries
		HD 3000plus and CPX 3000plus systems	3D Systems
		ProJet 6000	ProJet
		Axis 2.1 kit, Glider 3.0	BotMill
22	2011	Aerosol Jet print head, Objet260 Connex	Optomec
		Fortus 250mc	Stratasys
		Released AM File (AMF) format	AM Technologies
		RepRap based Buildatron 1 3D printer	Buildatron Systems
		ProJet 1500	3D Systems
		RepRap based Solidoodle 3D printer	Solidoodle
		Standard terminology for coordinate systems and test methodologies	ASTM International Committee F42 with ISO Technical Committee 261
		New line of ultrasonic additive manufacturing	Fabrisonic
		Online digital manufacturing and social networking facility released	Kraftwurx
		SLM 280 HL	SLM Solutions
		Freeform Pico	Asiga
		ProJet 1000	3D Systems
		LUMEX Avance-25	Matsuura
23	2012	MakerBot Replicator	MakerBot
		Cube	3D Systems
		MAGIC LF600	EasyClad
		II generation machine for $499	Solidoodle

(*Continued*)

TABLE 2.1 (*Continued*)

Timeline of AM Processes

Sr. No.	Year of Advent	Name of the Process/Modeler	Patented/Commercialized By
		Mojo 3D printer	Stratasys
		Objet30 Pro	Objet Geometries
		UP! Mini	Delta Micro Factory Corp.
		ProJet 7000	3D Systems
		EOS M series machines	EOS and Cookson Precious Metals
		Iris full-color, paper–based, sheet lamination system	MCor Technologies
		High-precision wax 3DP (3Z series)	Solidscape
		Flex Platform	ExOne
		Replicator 2 and 2X platforms	MakerBot Industries
		Matrix 300+ machine	Mcor
		Form 1	Formlabs
		Solidoodle 3D printer for $499	Solidoodle
		Objet1000 system	Objet Geometries
		VXC800 machine	Voxeljet
		Formiga P 110 laser sintering machine	EOS
		ProJet 3500 HDMax and ProJet 3500 CPXMax	3D Systems
		Ultimaker 3D printer	Ultimaker
24	2013	CubeX multihead 3D printer	3D Systems
		3Dent printer	Envisiontec
		Arcam Q10 machine	Arcam
		Replicator 2	MakerBot and Autodesk
		HeartPrint service	Materialise
		ProJet x60 family of machines	3D Systems
25	2014	Several new modelers and AM systems based on varied functionalities emerged from 2014 onward.	

Note: Despite the best efforts to list all the major processes and remarkable developments, the list may not be exhaustive owing to unreported international advances and the wide variety of these technologies.

2.3 Working Principles and Additive Manufacturing Process Chain

The process chain for AM is based on the general working principles of this technology and comprises six main steps, that is, obtaining a 3D computer aided design (CAD) model from different sources (Figure 2.2), tessellation, slicing, generation of scan/material deposition paths, generating physical models, and finishing/postprocessing [20,21]. Figure 2.3 illustrates the various steps involved in AM.

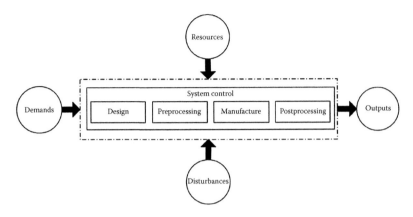

FIGURE 2.1
Concept of additive manufacturing industrial system. (From Eyers, D.R.a.P., Andrew, T. *Computers in Industry*, 2017. 92–93(Supplement C): 208–218.) [16]

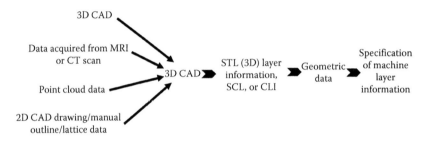

FIGURE 2.2
Line diagram for data flow in typical AM process.

2.4 Classification of Additive Manufacturing Techniques

There are many ways to classify AM techniques. The first is on the basis of the physical state of the bulk materials. The state may be liquid, powder or solid sheets, or gaseous. This is the most common way of classifying AM processes for prototyping requirements and is used by many researchers. The second basis of classification can be the mechanism employed to transfer sliced stereolithography (STL) data into physical structures. Following this, AM processes fall into four categories: one-dimensional channels, multiple one-dimensional channels, arrays of one-dimensional channels, and two-dimensional channels. A better way to streamline the classification is to combine both of these, that is, the raw material and its transfer mechanism approaches, as shown in Figure 2.4, which can be understood as third basis of classification. The empty boxes in this array can provide a foundation to researchers and industry personnel for further advancements in this

Additive Manufacturing Technologies

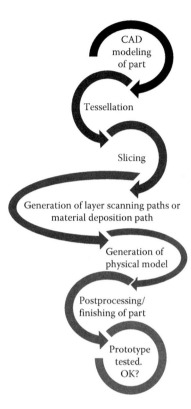

FIGURE 2.3
Flow chart of AM. (From Srivastava, M. *Some Studies on Layout of Generative Manufacturing Processes for Functional Components*, Delhi University, 2015.) [20]

		Mechanism of data transfer			
		One-dimensional	Multi one-dimensional	One-dimensional array	Two-dimensional
State of raw material	Liquid	Stereo lithography, liquid thermal polymerization		Objet quadra process	Solid ground curing, rapid microproduct development
	Discrete particles	Selective laser sintering, laser sintering technology, laser engineering net shaping, laser-assisted chemical vapor deposition, selective laser reactive sintering, gas phase deposition, selective area laser deposition	Laser sintering technology	Three-dimensional printing	Direct photo shaping
	Molten material	FDM, BPM, 3D welding, precision droplet based net-form manufacturing		Multijet modeling	Shape deposition modeling
	Solid sheets	Laminated object manufacturing, paper lamination technology			Solid foil polymerization
	Electroset fluids				Electro setting

FIGURE 2.4
Classification on the basis of bulk materials and data transfer mechanism. (From Srivastava, M. *Some Studies on Layout of Generative Manufacturing Processes for Functional Components*, Delhi University, 2015.) [20]

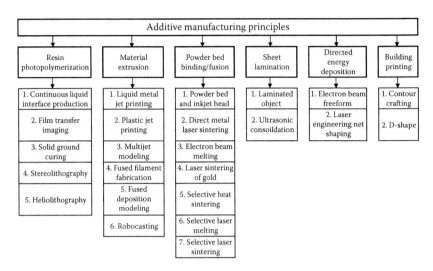

FIGURE 2.5
Classification of the basis of working principles.

field. The fourth basis of classification can be the working methodology or underlying technology. This is shown in Figure 2.5.

The fifth basis of classifying AM techniques is the methodology of component fabrication, under which there are two classes: fusion-based AM processes and solid-state AM processes. This is illustrated in Figure 2.6. The sixth basis of classification is the energy source or material-joining technology, for example, binders, lasers, electron beams, frictional heat, plasma arcs, and so on. The seventh basis of classification is the materials

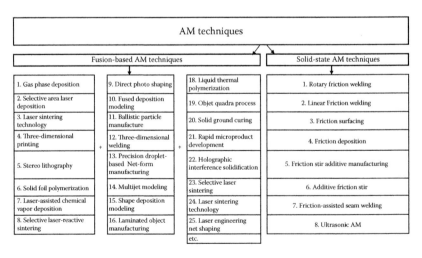

FIGURE 2.6
Classification on the basis of fabrication methodology.

TABLE 2.2

Classification on the Basis of Material Delivery System

Description	Powder Bed	Powder Feed
Deposition strategy	It uses a powder deposition system with a coating mechanism to spread layers of powder onto a substrate plate and powder reservoir.	Material flows through a nozzle being melted from a beam right on the surface of the treated part
Layer thickness	It produces 20–100-micron-size layers that can either be bonded together using lasers or stuck together using melting.	Highly precise systems based on layer deposition thickness between 0.1 mm to several centimeters.
Features	Choice of process depends upon: (a) laser unit, (b) powder handling, (c) built chamber size, etc.	Cladded substance has metallurgical bonding with the base material and undercutting is absent. The process is not same as other welding techniques.
Examples	(a) Selective laser sintering, (b) laser cusing, (c) direct metal laser sintering, (d) electron beam welding.	(a) Laser cladding, (b) laser metal deposition, (c) directed energy deposition, (d) laser-engineered net shaping.
Commercial systems	(a) Arcam AB, (b) Matsuura, (c) Hogonas.	(a) BeAM, (b) Trumpf, (c) Sciaky.

being used, for example, plastics, ceramics, powders, and so on. The eighth basis of classification is the material delivery system, as shown in Table 2.2. The ninth basis is the type of AM machine, that is, metallic or nonmetallic, as shown in Table 2.3.

A further description of individual techniques as per ASTM F42 Committee guidelines is the basis of the tenth classification system, which is presented in Table 2.4.

TABLE 2.3

Classification on the Basis of ASTM F42 Committee Guidelines (AM Machines)

Description	Metallic Machines	Nonmetallic Machines
Technology	Powder bed fusion, binder jetting, direct energy deposition, sheet lamination	Material extrusion, material jetting, vat photo-polymerization
Feedstock material	Metallic feedstock	Nonmetallic feedstock like powder, resins, plastics, glass, etc.
Physical state	Solid (powder, sheet, or wired form)	Solid (powder, sheet, wire), liquid or gas
State of consolidation	Consolidation of feedstock into parts of full density	Cannot produce full-density components

Source: ASTM International. *Standard Terminology for Additive Manufacturing Technologies: Designation F2792-12a* 2012, ASTM International: West Conshohocken, PA. [22]

TABLE 2.4

Description of Individual Techniques on the Basis of ASTM F42 Committee Guidelines (Materials and Technology)

S.No.	Technique	Description
1	VAT photo polymerization	Utilizes liquid photopolymer resin vat from which model is fabricated in a layered fashion.
2	Material jetting	Object creation takes place in a method identical to 2D ink jet printing in which material is jetted on a build platform utilizing a continuous or drop-on-demand technique.
3	Binder jetting	Selectively deposits a binder (in liquid form) on metal powder to join the powder particles. Fabrication is accomplished by a print head depositing layers.
4	Material extrusion	Drawing material from nozzle and its subsequent deposition takes place in layered form. Relative motion between nozzle and platform takes place after each layer.
5	Powder bed fusion	PBF includes those processes in which a focused energy source is utilized to selectively fuse/melt powder bed region. The energy sources in PBF mainly include laser or electron beams.
6	Sheet lamination	Includes processes like UAM, LOM, etc., and utilizes metallic sheet/ribbons mutually bounded via ultrasonic welding.
7	Directed energy	It covers laser-engineered net shaping (LENS), direct metal deposition (DMD), 3D laser cladding, etc., in a relatively complicated printing process normally utilized for repairing/adding material to existing parts.

Source: Loughborough University; http://www.lboro.ac.uk/research/amrg/about/ the7categoriesofadditivemanufacturing. [23]

2.5 Common Additive Manufacturing Processes

A list of some prominent AM processes is included in Table 2.5. This list is not exhaustive. It is merely an attempt to introduce some commercially important AM systems to the readers.

2.6 Advantages and Challenges of Additive Manufacturing Processes

Some of the most important advantages of AM include:

a. Is noise free
b. Can be operated from the comfort of home or office
c. Offers an excellent and impressive spectrum of applications

TABLE 2.5
Summary of Some Prominent AM Processes

S. No.	Technology	Raw Materials	Key Issues	Strengths	Applications
1	SLS	Either coated powders or powder mixtures, polymers, polystyrene, metals (steel, titanium, etc.), alloy mixtures, green sands	(a) Involves use of expensive and potentially dangerous lasers. (b) Components have porous surfaces.	(a) Fully self-supporting, so nesting is very easy (b) Parts offer high strength, stiffness, and chemical resistance	Physical models of aircrafts, used especially where a small number of high-quality parts is required
2	SGC	Photosensitive resins	Produces large amount of waste, high operating cost, complex system.	Does not require support structure (as wax fills the void), good accuracy, high fabrication rates	Models, prototypes, patterns, and production parts
3	Inkjet printing (IJP)	Wax-like thermoplastic materials	Ink is expensive, precise control of jets is required, fading with time is predominant in components.	Clean and low-priced concept modelers Suited for small, accurate, and intricate parts	Suitable for concept modeling and investment casting models
4	FDM	Thermoplastics, nylon, ceramics, metals, etc.	Slow speed, leading to high build time. Anisotropy, inferior surface finish.	Cost, material	Form and functional models, concept models, prototypes, medical models
5	SLA	Liquid resins that can be hardened using photo polymerizing effect	Process controlled by laser quality, scanning periods, resin recoating; requires support structures; portability issues due to liquids, postprocessing problems.	One of the oldest AM techniques, underwent lots of research and emphasis during initial phases, sufficiently large components can be made accurately	Now obsolete chiefly owing to the need for expensive resins and lasers

(Continued)

TABLE 2.5 (Continued)
Summary of Some Prominent AM Processes

S. No.	Technology	Raw Materials	Key Issues	Strengths	Applications
6	LOM	Paper, wood, plastic sheets	Dimensional accuracy less than SLA and SLS, limited scope of material, surface finish and accuracy issues.	(a) No milling necessary (b) Low cost due to readily available raw materials (c) Large components can be handled	Rapid tooling, prototyping, pattern making, medical applications
7	LENS	Metallic powder of steels, nickel-based super alloys, Inconel, titanium, cobalt	Severe overhangs, solidification microstructures, inferior mechanical properties, surface finish, secondary finishing process required, support structure problems.	Can use composite powder mixture, lowers cost and time requirements, parts fully dense with nondegraded microstructures	Repair, overhaul, production for aerospace, defense and medical markets
8	Laser additive manufacturing (LAM)	Steels, nickel-based super alloys, Inconel, titanium, cobalt	(a) Severe overhangs, (b) solidification microstructures, (c) inferior mechanical properties.	(a) Can use composite powder mixture, (b) High cooling rates	Aerospace, defense, automotive, and biomedical industries
9	Laser sintering	Multicomponent powders composed of high melting point components called structural metal, low melting point components called binder and flux/deoxidizing agents called additives	(a) Inert gas environment required. (b) High-quality laser systems are required that also highly affect powder consolidation. (c) Mushy zone due to solid liquid wetting.	(a) Can produce multimaterial alloys (b) Appreciably good mechanical properties can be obtained by controlling process parameters	Medical models, metallic molds, etc.

(*Continued*)

TABLE 2.5 (Continued)
Summary of Some Prominent AM Processes

S. No.	Technology	Raw Materials	Key Issues	Strengths	Applications
10	Laser melting	Multimaterial ferrous and nonferrous powders	(a) High-quality laser systems are required. (b) Requires high energy levels. (c) Instability of melting points need attention.	(a) Produces fully dense components (b) Mechanical characteristics comparable to bulk materials (c) Can process nonferrous materials	Pure nonferrous metal full-density components with high strength like titanium, multimaterial components, etc.
11	Laser metal deposition	Multimaterial powders	(a) Requires high-quality specially designed coaxial feeder system. (b) Requires high-quality laser systems. (c) Costly and intricate. (d) Requires intricate closed-loop control.	(a) Quite advanced (b) Multimaterial delivering capabilities (c) Patented closed loop controls	Manufacturing new components for repair and rebuilding of worn out/damaged products, for wear and corrosion resistant coatings, etc.
12	Digital metallization	Metal foil, plastics, paper	(a) Expensive. (b) Few modelers available. (c) Is suitable for late-stage customization.	(a) Can be used to produce design of different colors and light effects (b) Suitable as a cost-effective customization technology to implement metallic effects	Decorative items, functional items like automobiles, electromagnetic shielding, circuit paths, etc.

(Continued)

TABLE 2.5 (*Continued*)
Summary of Some Prominent AM Processes

S. No.	Technology	Raw Materials	Key Issues	Strengths	Applications
13	Prometal	Iron, bronze, glass, etc.	(a) Low speeds. (b) Limited volumes. (c) Technological and economical limitations.	(a) Offers customized hardware in very little time (b) Suitable for both metals and glasses	3D printed door pulls, knobs, knockers, decorative items, customized 3D hardware components
14	DMD	Steel, wasp alloy, titanium	(a) Postprocessing. (b) Low build rate. (c) Complex process.	(a) Neutral gas, (b) precision part pickup (c) quick tool path operation	Used to repair and rebuild worn/damaged parts, tools, dies, cutters, etc.
15	SLM	Steel, Inconel, titanium, cobalt, aluminum	Postprocessing, high production cost, properties.	Complex geometry is achievable	Medical and dental applications, lightweight structures, heat exchangers
16	Easy clad	Steel, wasp alloy, titanium	Postprocessing, low build rates.	Neutral gas, better properties in comparison to castings	Construction of parts for aerospace and aircraft industries
17	Ultrasonic additive manufacturing (UAM)	Plastic or metallic sheet like aluminum alloys	Different microstructure at interfaces and noninterfaces. Weak links in build, foil preparation, low build volume.	Solid-state processing method; multimaterial structures; environmentally friendly; involves less heat so no melting, hence no distortion; good bonding properties	Injection molding dies, parts with embedded channels, incorporation of second-phase materials, intricate geometry components, etc.

(*Continued*)

Additive Manufacturing Technologies

TABLE 2.5 (Continued)
Summary of Some Prominent AM Processes

S. No.	Technology	Raw Materials	Key Issues	Strengths	Applications
18	Direct metal laser sintering	Steels, Al, cobalt, titanium	Postprocessing. High operating cost.	Multimaterial structures	Turbine blades
19	Laser cusing	Precious metals, steels, titanium, aluminum	Mechanical properties, low build rates.	High-quality finishing, reduction in stresses	Dental, jewelry automotive, aerospace, medical devices, tool and mold manufacture
20	Electron beam melting (EBM)	Fine metallic powders of copper, beryllium, steels, titanium, aluminum, nickel, etc., and alloys	Surface quality, solidification defects, high skill requirement, large power requirement.	Faster build compared to DMLS and SLM; delivers mechanical strength with less mass and cost, and weight is reduced	Small series parts, prototypes, support parts (jigs, fixtures, etc.)
21	Shaped metal deposition	Metallic wires	Poor dimensional accuracy and surface finish, environmental hazards, high cost.	High production rates, lower costs, shorter lead times, high deposition rates and efficiencies, denser part production capability	Features with intricacy and complexity, large-scaled parts (e.g., aerospace and metal die)
22	Electro-discharge deposition	Any material that is conductive can be utilized as raw material	Magnetic field needs to be very carefully controlled, as it has major impact over deposition.	(a) Used to manufacture microproducts (b) Integration of multitude of functions is possible	Miniatured products with high aspect ratios, microfeatures
23	Selective layer chemical vapor deposition	Metals like aluminum, copper, tungsten, etc.	(a) Controlling chemistry of vapor. (b) Choosing a suitable vapor deposition technique.	(a) Conformal coverage of irregularly shaped surfaces (b) High throughput	Thin film applications, functionally gradient materials
24	Shape deposition modeling	Powder (stainless steel invar, aluminum, titanium)	High cost, support material requires manual work for removal.	Wide variety of material; use of conventional machine tools; no stair-step effect, unlike others	Dense components with accuracy, excellent surface finish and metallurgical bonding

d. Can form process chains when suitably combined with other conventional/unconventional manufacturing processes
e. Offers considerable reduction in lead times in terms of both component manufacturing and time to reach market
f. Provides considerable reduction in cost of machining, and material removal wastage is estimated at around 80% over conventional techniques
g. Is an established prototyping tool for high-quality prototypes
h. Creates parts with complex and intricate geometries
i. Has an absence of tools, jigs, fixtures, molds, punches, and so on
j. Offers a remarkable reduction in human intervention

However, apart from the benefits stated above, AM faces certain limitations that have restrained its acceptance as a full-scale commercial technology, especially in application areas where strength is an important requisite. A non-process-specific list of the major challenges of AM is below:

a. Nonoptimal build speeds
b. Comparatively lower accuracy
c. Decisions needed regarding optimal part orientation
d. Restricted choice of raw materials and resulting material properties
e. Poor surface finish
f. Preprocessing and postprocessing requirements
g. Expensive process cost owing to limited customers
h. Anisotropic behavior of AM-fabricated parts
i. Occurrence of stair-stepping phenomenon
j. Layout optimization required for each process
k. Use of support structures
l. Poor structural strength of parts fabricated via AM techniques
m. Consistently changing and upgrading AM technologies

2.7 Applications of Additive Manufacturing Technologies

AM offers an impressive spectrum of applications. In fact, there is perhaps no sphere that is untouched by AM. It offers mould free processes that are highly suitable for making intricate geometrical models. The adaptable flexibilities of AM techniques pose it as a successful potential candidate for making complex biomedical models. This feature additionally makes AM an excellent choice for numerous other highly customized and complex geometrical applications.

Initially, there were six major classes of components fabricated via AM processes, which included functional prototypes, concept models, a few functional parts and tools, micromachining, functionally graded materials, and medical parts. Today, the application of components manufactured using AM has extended to almost every practical aspect, including manufacturing, materials, medicine, defense and commercial applications, and so on. This can be seen in the exceedingly large number of patents being filed each year in this field. Drives by giants like GE, Boeing, Airbus, and so on vouch for the facts related to economic growth in the field of AM [24–27]. Economically, around 1995, sales growth was 40%–50%. In 1999, it was around 22%. AM was rightly predicted to have enormous growth and possess an asset size of $3 billion by 2016. The investment of $50 million by GE for a jet engine nozzle presents a picture of the commercial success of AM techniques. Major AM-enabled projects have increased in the transportation industry. A lot of fuel saving can be accomplished by using parts fabricated via this route [28,29]. Market researchers have predicted generation of a worldwide $200–$600 billion annual economic market for AM technologies by 2025 [30]. The use of MAM for lightweight transportation industries would considerably optimize result, but AM's share in the overall transport and aviation industries is still very small owing to three major reasons: (a) structural incompetence, (b) raw material cost, and (c) limitations on the number of alloys that can be fabricated.

A list of various major industries, along with reasons for the success of AM technology in the specific sector, is given below:

1. Aircraft and Automobile Industry: The applications of AM are increasing in the aerospace and automobile sectors, chiefly because of reduced design and development times, the ability to fabricate metallic prototypes from metals like titanium, the ability to create complex and improved-performance products, the ease of making lightweight parts, capabilities for product design verification, and so on. With the use of AM-fabricated parts, considerable material savings can be achieved. For example, Petrovic et al. [31] concluded in their work that material savings of up to 40% can be achieved by using AM as compared to conventional techniques.

    The advantages of utilizing AM in the aerospace sector include reduced lead time to introduce products by up to 30% to 70%, reduced nonrecurring costs to introduce products by up to 45%, and reduced cost to manufacture low-volume service components by 30% to 35%.

2. Artistic Industry: Intricate and complex jewelry can be developed via AM owing to high printing resolution. These techniques are

powerful tools for artists and jewelers. SLA, Envisiontec GmbH, and the Solidscape division of Stratasys have shown keen interest in the application of AM techniques for such uses.

3. Medical Industry: AM technologies are being successfully used in medical industry, tissue engineering, and biomedical applications. They provide good visualization aid in presurgical planning, ease rehearsals for surgical procedures, and can act as communication interfaces. Various AM techniques are utilized to fabricate biomedical components, create tailored hearing devices, place implants, create bone scaffolds, help with dental basics and complex surgeries, reconstruct bone plates and bone grafts, and so on. A few AM applications are presented in Figure 2.7. However, significant future research is required in this field to successfully harness its complete potential.

4. Architectural Industry: Another interesting AM application is in the architecture industry, where it helps save time by eliminating the need for difficult handmade models. The capability to create complex shapes precisely saves a lot of design time. Many extremely intricate shapes and their further refinements can be obtained at the design stage itself [31].

Other major industries that highly benefit from AM techniques include:

5. Retail/apparel
6. Machine tooling
7. Food
8. Customized goods

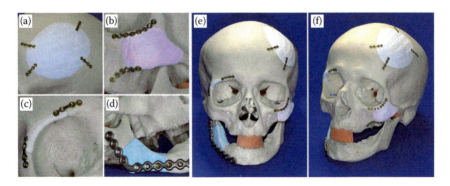

FIGURE 2.7
(a–f) Various additively manufactured implants. (From Klammert, U. et al. *Journal of Cranio-Maxillofacial Surgery*, 2010. 38: 565–570.) [32]

2.8 Metal Additive Manufacturing Techniques

Today, there are more than a hundred AM methods. This is mainly owing to immense growth in the field of materials, communication strategies, CAD/CAM (computer aided manufacturing), and so on. Owing to increasing demands from various critical industries like aerospace, biomedical, medical, tooling, and so on, considerable research these days is focused on MAM. The use of AM techniques for the production of metallic parts is referred to as MAM. The suitability of MAM techniques as compared to conventional manufacturing techniques is mainly due to their ability to fabricate complex geometrical metallic components in relatively less time.

Among the preliminary reports on MAM setups was a machine utilizing lasers for selective melting of metallic layers. The first metallic 3D part was fabricated from Cu, Sn, and Pb-Sn solder in 1990 [33]. Of the seven categories described in Table 2.4, four categories are based on MAM, that is, powder bed fusion (DMLS, SLS, and SLM [selective laser melting]), binder jetting (3DP), direct energy deposition (laser-engineered net shaping [LENS], Laser Cusing [LC], electron beam freeform fabrication [EBF³], and directed light fabrication [DLF]) and sheet lamination (laminated object manufacturing [LOM], ultrasonic consolidation [UC], etc.) [22,32–35]. The remaining three categories are currently not applicable to MAM [35]. The chronological

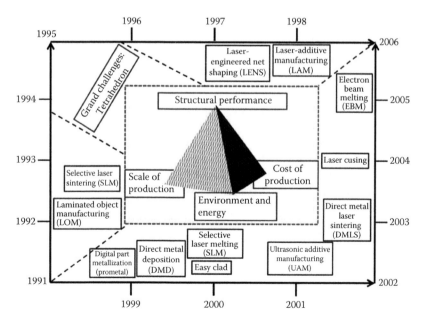

FIGURE 2.8
Chronological development of MAM techniques and challenges associated with them. (From Palanivel, S., Sidhar, H., Mishra, R. S. *JOM*, 2015. 67(3): 616–621.) [36]

TABLE 2.6

Comparison of Some Prominent MAM Processes

Additive Materials	Process	Layer Thickness (μm)	Deposition Rate (g/min)	Dimensional Accuracy (mm)	Surface Roughness (μm)	References
Powder	LC	N/A	1–30	±0.025–±0.069	1–2	[37]
	SLM	20–100	N/A	±0.04	9–10	[38,39]
	SLS	75	~0.1	±0.05	14–16	[40]
	DLF	200	10	±0.13	~20	[41]
Wire	WAAM	~1500	12	±0.2	200	[42]
	EBF[3]	N/A	Up to 330	Low	High	[43]

Source: Ding, D. et al. *The International Journal of Advanced Manufacturing Technology*, 2015. 81(1): 465–481. [34]

development of these MAM techniques is presented in Figure 2.8 [36], and a general comparison of these processes is presented in Table 2.6.

The MAM processes that belong to specific categories are described below.

The benefits, key issues, applications, and raw materials for each of these are discussed in detail in Table 2.5.

2.8.1 Limitations of Metal Additive Manufacturing

The major available MAM techniques are based on powder bed fusion and direct energy deposition principles. Owing to the involvement of fusion/molten pools, these techniques suffer from various limitations in the form of metallurgical defects [44]. Some of the main defects occurring during MAM techniques are discussed here.

2.8.2 Porosity

Porosity and voids are prominent MAM defects that should be eliminated because they adversely affect properties of fabricated builds [45–47]. These can occur owing to one of the following mechanisms: (a) repeated formation and collapse of unstable keyholes, amounting to voids of entrapped vapor in the deposit. Keyhole mode to melt/deposit is normally adopted when AM processes must run at appreciably high-power density [48–50]. (b) Gas entrapment and the resulting microscopic pores of spherical shape in powder particles when it is atomized. Sometimes, shielding gas or alloy vapor can also be entrapped. (c) If the melt pool of the next layer does not adequately penetrate into either the substrate or preceding layer, then pores can be formed. Keyholes porosity, lack of fusion pores, and gas-induced porosity defects are shown in Figure 2.9 [48,51].

Additive Manufacturing Technologies

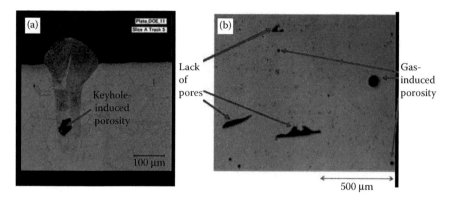

FIGURE 2.9
Defects during MAM: (a) keyhole porosity. (From King, W. E. et al. *Journal of Materials Processing Technology*, 2014. 214(12): 2915–2925.) [48]; (b) pores owing to lack of fusion and gas-induced porosity. (From Sames, W. et al. Effect of process control and powder quality on Inconel 718 produced using electron beam melting, in *Proceedings of the 8th International Symposium on Superalloy 718 and Derivatives*; 2014. p. 409.) [51]

2.8.3 Loss of Alloying Elements

Alloying elements vaporize during AM owing to appreciably high temperatures of the melt pool. Owing to the variable volatility of a few elements over others, these are selectively vaporized, which leads to appreciable changes in overall alloy composition [44]. Compositional changes in turn influence solidification microstructure, corrosive behavior, and mechanical characteristics. This can cause severe issues, especially during fabrication of high-quality critical parts. It is expected that the most important AM process parameters in the vaporization of alloys include power, scanning speed, feedstock feeding rates, and beam dimensions.

2.8.4 Cracking and Delamination

A few different kinds of cracking can occur in AM parts [52]. The first is solidification-based welding like cracking [53]. This is prominent along grain boundaries and can mainly be attributed to temperature gradients owing to differential contraction rates of different layers (solidifying layer, depositing layer, and substrate) [54]. The second is liquation cracking [53], chiefly seen in mushy/partially melted zones of parts that are subjected to tensile force under which liquid films act as cracking points [54]. Other types of cracks can be similarly understood. AM cracks can either be substantially long, as shown in Figure 2.10a, or relatively small, as shown in Figure 2.10b [55]. Two layers are said to be delaminated if they are mutually separated. This happens if the residual stress at the layer interface exceeds the alloy's yield strength [56].

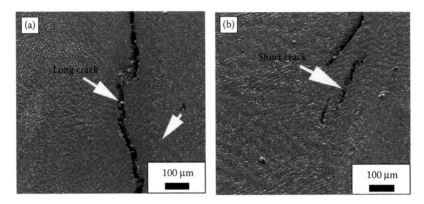

FIGURE 2.10
Cracks in Rene88DT superalloy fabricated via laser solid forming: (a) long crack, (b) short crack. (From Zhao, X. M. et al. *Materials Science and Engineering A* 2009. 504(1–2), 129–134.) [55]

Thus, commonly used MAM techniques are prone to defect formation owing to fusion. Common applications of MAM techniques based on fusion are in manufacturing components for biomedical implants; functionally graded and refractory materials; and components for space, automotive, marine and other transportation sectors, and so on.

2.9 Conclusion and Future Scope of Additive Manufacturing

AM technologies are witnessing a transition from prototyping to end tooling and manufacturing applications. Considerable progress in the processes has brought about the ability to produce accurate multimaterial parts in reduced times. Currently, these technologies are quite proficient in fabricating parts with superior density and mechanical properties. AM possesses varied advantages above conventional manufacturing techniques, as discussed in the above sections, in industries like aircraft, automobile, architecture, art, and medicine (tissue engineering and biomedical applications). Despite all these favorable features, full-scale utilization is still a challenge owing to manifold limitations and challenges, as discussed in the above section. These issues also constrain the utilization of AM in certain critical sectors requiring reduced costs with improved precision. In order to optimize a given AM process, various qualitative and quantitative responses for specific processes like build parameters, local compositional control, layout control, communication interface controls, mechanical and physical property control, environmental and energy considerations, and economic cost benefit analysis must also be optimized.

Much of the research's scope lies untapped in the domains of reducing overall cost, developing newer materials, developing processes, and optimizing performance via the AM route for customized applications. Owing to the specific advantages of material saving, ease of fabrication, and so on, considerable research these days is focused on MAM. Efforts are in progress to commercialize methods of coupling advancements in conventional manufacturing to the features of AM technologies. Processes like friction based AM techniques are emerging and can open altogether new avenues in the enhancement of the material microstructures and mechanical properties of the fabricated functional components.

References

1. Pham, D.T., Demov, S.S., *Rapid Manufacturing: The Technologies and Applications of Rapid Prototyping and Rapid Tooling* 2001. Springer-Verlag: London.
2. Pham, D.T., Dimov, S.S., Rapid prototyping and rapid tooling—The key enablers for rapid manufacturing. *Proceedings of the Institution of Mechanical Engineering Science: Journal of Mechanical Engineering Science: SAGE Journals*, 2003. 217: 3–24.
3. Berman, B., 3-D printing: The new industrial revolution. *Business Horizons*, 2012. 55(2): 155–162.
4. Hopkinson, N., Hague, R., Dickens, P., *Rapid Manufacturing: An Industrial Revolution for the Digital Age* 2006. John Wiley & Sons: Chichester, England.
5. ASTM F2792-10 *Standard Terminology for Additive Manufacturing Technologies* 2010, ASTM International: West Conshohocken, PA.
6. Pham, D.T., Gault, R.S., A comparison of rapid prototyping technologies. *International Journal of Machine Tools and Manufacture*, 1998. 38(10–11): 1257–1287.
7. Kruth, J.P., Leu, M.C., Nakagawa, T., Progress in additive manufacturing and rapid prototyping. *CIRP Annals-Manufacturing Technology*, 1998. 47(2): 525–540.
8. Kruth, J.-P., Material incress manufacturing by rapid prototyping techniques. *CIRP Annals*, 1991. 40(2): 603–614.
9. Gebhardt, A., *Understanding Additive Manufacturing: Rapid Prototyping, Rapid Tooling, Rapid Manufacturing* 2012. Hanser Publications: Cincinnati.
10. Wohlers, T., Gornet, T., History of additive manufacturing. *Wohlers Report 2014 - 3D Printing and Additive Manufacturing State of the Industry*, 2014. 24: 1–34.
11. Printing Industry, History of 3D Printing 2017. [Online]. Available: https://3dprintingindustry.com/3d-printing-basics-free-beginners-guide/history/
12. Wohlers, T., Wohlers Report 2000: Executive summary. *Time-Compression Technology*, 2000. 8(4): 29–31.
13. Wohlers, T., *Additive Manufacturing and 3D Printing State of the Industry*, 2012. Report, 1–287.
14. Wikimedia foundation Inc., Wikipedia—Powder bed and inkjet head 3D printing technologies. [Online]. Available: https://en.wikipedia.org/wiki/Powder_bed_and_inkjet_head_3D_printing
15. Campbell, I., Bourell, D., Gibson, I., Additive manufacturing: Rapid prototyping comes of age. *Rapid Prototyping Journal*, 2012. 18(4): 255–258.

16. Eyers, D.R.a.P., Andrew, T., Industrial additive manufacturing: A manufacturing systems perspective. *Computers in Industry*, 2017. 9293(Supplement C): 208–218.
17. Attaran, M., The rise of 3-D printing: The advantages of additive manufacturing over traditional manufacturing. *Business Horizons*, 2017. 60(5): 677–688.
18. Guo, N. Leu, M. C., Additive manufacturing: Technology, applications and research needs, Front. *Mechanical Engineering*. 2013, 8(3): 215–243.
19. Bose, S., Ke, D., Sahasrabudhe, H., Bandyopadhyay, A., Additive manufacturing of biomaterials. *Progress in Materials Science*, 2018. 93: 45–111.
20. Srivastava, M., *Some Studies on Layout of Generative Manufacturing Processes for Functional Components*. Delhi University, 2015.
21. Pandey, P.M., Rapid prototyping technologies, applications and Part deposition planning, 2012. [Online]. Available: iitd.ac.in/~pmpandey/MEL120_html/RP_document.pdf.
22. ASTM International. *Standard Terminology for Additive Manufacturing Technologies: Designation F2792-12a2012*. ASTM International: West Conshohocken, PA.
23. Loughborough University official website; http://www.lboro.ac.uk/research/amrg/;2017
24. Koykendall, J., Cotteleer, M., Holdowsky, J., Mahto, M., *Report on 3D Opportunity for Aerospace and Defense: Additive Manufacturing Takes Flight*, 2014. https://www2.deloitte.com/insights/us/en/focus/3d-opportunity/additive-manufacturing-3d-opportunity-in-aerospace.html?ind=70
25. Beaman, J.J., Solid Freeform Fabrication, US Patent No. 473901, DOI 10.1109/6.7448741892. 584–595.
26. Kruth, J.P., Leu, M.C., Nakagawa, T., Progress in additive manufacturing and rapid prototyping. *CIRP Annals-Manufacturing Technology*, 1998. 47(2): 525–540.
27. Lyons, B., In: Additive manufacturing in aerospace; examples and research outlook. *The Boeing Company*, September 2011 https://www.naefrontiers.org/File.aspx?id=31590.
28. Luo, Y., Ji, Z., Leu, M.C., Caudill, R., Environmental performance analysis of solid freeform fabrication processes, in *Proceedings of the 1999 IEEE International Symposium on Electronics and the Environment (Cat. No.99CH36357)* 1999, 1–6.
29. Drizo, A., Pegna, J., Environmental impacts of rapid prototyping: An overview of research to date. *Rapid Prototyping Journal.*, 2006. 12(2), 64–71.
30. Manyika, J., Chui, M., Bughin, J., Dobbs, R., Bisson, P., Marrs, A., *Disruptive Technologies: Advances That Will Transform Life, Business, and the Global Economy* 2013. McKinsey Global Institute: Seoul, Korea and San Francisco, US.
31. Petrovic, V., Haro, J.V., Jordá, O., Delgado, J., Blasco, J.R., Portoles, L., Additive layer manufacturing: State of the art in industrial applications through case studies. *International Journal of Production Research, Taylor & Francis*, 2010. 1: 1061–1079.
32. Klammert, U., Gbureck, U., Vorndran, E., Rödiger, J., Meyer-Marcotty, P., Kübler, A.C., 3D powder printed calcium phosphate implants for reconstruction of cranial and maxillofacial defects. *Journal of Cranio-Maxillofacial Surgery*, 2010. 38: 565–570.
33. Frazier, W.E., Digital manufacturing of metallic components: Vision and roadmap, in *Solid Free Form Fabrication Proceedings*, 2010, 717–732.
34. Ding, D., Pan, Z., Cuiuri, D., Li, H., Wire-feed additive manufacturing of metal components: Technologies, developments and future interests. *The International Journal of Advanced Manufacturing Technology*, 2015. 81(1): 465–481.

35. Sames, W.J., List, F.A., Pannala, S., Dehoff, R.R., Babu, S.S., The metallurgy and processing science of metal additive manufacturing. *International Materials Reviews*, 2016. 61(5): 315–360.
36. Palanivel, S., Sidhar, H., Mishra, R.S., Friction stir additive manufacturing: Route to high structural performance. *JOM*, 2015. 67(3): 616–621.
37. Xue, L., Islam, M.U., Laser consolidation-a novel one-step manufacturing process for making net-shape functional components, in *Cost Effective Manufacturing via Net-Shape Processing* 2006, 15-1–15-14. Neuily-sur-Seine, France: Meeting Proceedings RTO-MP-AVT-139. http://www.rto.nato.int/abstracts.asp.
38. Mumtaz, K., Hopkinson, N. Top surface and side roughness of Inconel 625 parts processed using selective laser melting. *Rapid Prototyping Journal*, 2009. 15(2): 96–103.
39. Mumtaz, K.A., Hopkinson, N., Selective Laser Melting of thin wall parts using pulse shaping. *Journal of Materials Processing Technology*, 2010. 210(2): 279–287.
40. Zhu, H.H., Lu, L., Fuh, J.Y.H., Development and characterisation of direct laser sintering Cu-based metal powder. *Journal of Materials Processing Technology*, 2003. 140(1): 314–317.
41. Milewski, J.O., Lewis, G.K., Thoma, D.J., Keel, G.I., Nemec, R.B., Reinert, R.A., Directed light fabrication of a solid metal hemisphere using 5-axis powder deposition. *Journal of Materials Processing Technology*, 1998. 75(1): 165–172.
42. Colegrove P., High deposition rate high quality metal additive manufacture using wire + arc technology 2010. https://www.xyzist.com/wp-content/uploads/2013/12/Paul-Colegrove-Cranfield-Additive-manufacturing.pdf.
43. Taminger, K.M., Hafley, R.A., Electron beam freeform fabrication for cost effective near-net shape manufacturing, 2006. https://ntrs.nasa.gov/archive/nasa/casi.ntrs.nasa.gov/20060009152.pdf
44. DebRoy, T., Wei, H.L., Zuback, J.S., Mukherjee, T., Elmer, J.W., Milewski, J.O., Beese, A.M., Wilson-Heid, A., De, A., Zhang, W., Additive manufacturing of metallic components – process, structure and properties. *Progress in Materials Science*, 2018. 92: 112–224.
45. Jia, Q., Gu, D., Selective laser melting additive manufacturing of Inconel 718 superalloy parts: Densification, microstructure and properties. *Journal of Alloys and Compounds*, 2014. 585(Supplement C): 713–721.
46. Morgan, R., Sutcliffe, C.J., O'Neill, W., Density analysis of direct metal laser re-melted 316L stainless steel cubic primitives. *Journal of Materials Science*, 2004. 39(4): 1195–1205.
47. Carlton, H.D., Haboub, A., Gallegos, G.F., Parkinson, D.Y., MacDowell, A.A., Damage evolution and failure mechanisms in additively manufactured stainless steel. *Materials Science and Engineering: A*, 2016. 651(Supplement C): 406–414.
48. King, W.E., Barth, H.D., Castillo, V.M., Gallegos, G.F., Gibbs, J.W., Hahn, D.E. et al. Observation of keyhole-mode laser melting in laser powder-bed fusion additive manufacturing. *Journal of Materials Processing Technology* 2014. 214(12): 2915–2925.
49. Svensson, M., Ackelid, U., Ab, A., Titanium alloys manufactured with electron beam melting mechanical and chemical properties, in *Proceedings of the Materials and Processes for Medical Devices Conference* 2010. 189–194.
50. Darvish, K., Chen, Z.W., Pasang, T., Reducing lack of fusion during selective laser melting of CoCrMo alloy: Effect of laser power on geometrical features of tracks. *Materials & Design*, 2016. 112: 357–366.

51. Sames, W., Medina, F., Peter, W., Babu, S., Dehoff, R., Effect of process control and powder quality on Inconel 718 produced using electron beam melting, in *Proceedings of the 8th International Symposium on Superalloy 718 and Derivatives* 2014. 409.
52. Kempen, K., Thijs, L., Vrancken, B., Buls, S., Van Humbeeck J., Kruth, J.. Producing crack-free, high density M2 Hss parts by selective laser melting: Pre-heating the baseplate, in *Proceedings of the 24th International Solid Freeform Fabrication Symposium; Laboratory for Freeform Fabrication*, Austin, TX, 2013. 131–139.
53. Carter, L.N., Attallah, M.M., Reed, R.C., Laser powder bed fabrication of nickel-base superalloys: Influence of parameters; characterisation, quantification and mitigation of cracking. *Superalloys* 2012. 2012: 577–586.
54. Kou, S., *Welding Metallurgy*. 2nd ed. John Wiley & Sons: Hoboken, NJ, 2003.
55. Zhao, X.M., Lin, X., Chen, J., Xue, L., Huang, W.D., The effect of hot isostatic pressing on crack healing, microstructure, mechanical properties of Rene88DT superalloy prepared by laser solid forming. *Materials Science and Engineering A* 2009. 504(1–2): 129–134.
56. Mukherjee, T., Zhang, W., DebRoy, T., An improved prediction of residual stresses and distortion in additive manufacturing. *Computational Materials Science* 2017. 126: 360–372.

3
Friction Based Joining Techniques

3.1 Introduction

Friction welding (FW) is a solid-state process of joining materials (typically suited for similar or dissimilar welding) with the application of frictional heat and pressure. The frictional heat is generated at the interface of components (having relative motion) that rub against each other under pressure. Material softens under heat, and the softened material consolidates to initiate the formation of the weld. The rubbing action is stopped when enough heat and plastic deformation is obtained. However, pressure is still applied for some time in the stationary state to allow development of the weld in the solid state. The American Welding Society (AWS) defines FW as a solid state joining method in which materials coalesce under pressure. The heat of friction at the contact surfaces generated owing to the relative movement of surfaces with respect to each other under a normally applied force is responsible for raising temperatures equivalent to forging conditions. Thus, FW is based on the conversion of mechanical energy into heat energy for joining two parts. The parent material does not melt during the course of FW execution, and solid-phase metallurgical bonding is obtained. This is in complete contrast to fusion welding techniques that generally require melt pool generation [1].

This chapter outlines the historical developments of FW. An overview of different FW variants based upon process-specific functionality is then presented in brief. This is followed by a brief discussion on the development of solid-state friction based additive manufacturing techniques. A timeline of developments in the domain of these FATs is then presented. Finally, the chapter concludes after a discussion of the benefits and limitations of this innovative hybrid manufacturing approach.

3.2 Historical Development of Friction Welding

Utilization of friction as a high performance thermomechanical source for welding and processing of materials has witnessed tremendous progress since the initial patent filed by Bevington in 1891 [2]. However, at that time, not much attention was paid to this novel idea. The utilization of FW came into existence in 1960 when Bishop summarized various aspects of FW [3]. Up to that time, FW was utilized at the laboratory scale. After this, the industrial application of FW started, and numerous developments and research have been carried out since then. Initially, rotary FW was invented, and it is the oldest FW technique. However, there is an inherent limitation of RFW because of its unsuitability in welding of noncircular parts. At least one part should be circular in RFW. This limitation was first addressed by German Richter in 1929 [4] and then by Vill [5] in 1962 by introducing linear friction welding. LFW allows welding of noncircular objects. In this direction, the Caterpillar Tractor Company patented apparatus for linear (reciprocating) friction welding in 1969 [6]. Another version of rotary friction welding, that is, inertia friction welding (IFW), came into existence in the early 1960s [1,7]. Various strategies of using the principles of frictional energy emerged during the 1980 s. In 1991, the invention of friction stir welding was a splendid feather in the cap of the friction welding area [8]. These techniques are still progressing, and numerous hybrid FW techniques are evolving.

3.3 Friction Welding Techniques

Friction welding techniques (FWTs) constitute an excellent class of productive, effective, and controllable fabrication and processing techniques to plasticize a given material area, thereby easily removing contaminants from the welded zone. They are economical and cost-effective green technologies that are utilized to fabricate components with welds of high structural strength.

Following are the significant advantages of friction welding techniques:

1. Green, environment friendly processes
2. No melting or related defects involved
3. No fumes, gas, or smoke are generated
4. Flux or gas for shielding purposes is not required
5. Filler material is not needed
6. Thermal energy is effectively utilized
7. Similar/dissimilar welding feasible
8. Ideal process to join materials with vast differences in melting points and mechanical characteristics
9. Can weld otherwise normally incompatible components
10. Less joint preparation time

11. Consistent phenomenon that is repetitive, automation compatible, and needs less reliance on skill
12. Can be used on any scale (conceptual to production)
13. No porosity or slag included; thus, defect-free joints are obtained
14. Better microstructures obtained
15. Thinner heat-affected zones obtained
16. Highly efficient and fast processes

These FWTs can be categorized on the basis of relative motion between the two parts.

3.4 Variants of Friction Welding Techniques

In a broader sense, there are three variants of FW: rotary, linear, and orbital FW [1]. However, there are various derived friction based processes. In 2003, Nicholas introduced and reviewed 16 friction based processes [9]. These processes can be listed as:

1. Rotary FW [5,10]
    a. Continuous drive FW (CDFW)
    b. Inertia FW (IFW)
2. Linear and angular FW (AFW) [11]
3. Friction surfacing [9]
4. Friction transformation hardening [12]
5. Friction stir welding [8]
6. Friction seam welding (Klopstock) [13]
7. Friction seam welding (LUC) [14]
8. Friction taper stitch welding [15,16]
9. Radial FW [17,18]
10. Orbital FW [19]
11. Friction hydropillar processing [20]
12. Friction extrusion and friction co-extrusion cladding [21]
13. Friction stud welding
14. Friction brazing [22]
15. Third-body FW [23]
16. Friction plunge [24]

Continued developments in the area facilitate the evolution of new variants. The following are more variants of friction welding that have emerged:

1. Friction deposition
2. Friction stir processing
3. Friction stir additive manufacturing
4. Friction assisted seam welding
5. Additive friction stir

Owing to constraints of space and scope of the present book, a detailed discussion of these friction based processes is not presented here. However, a brief introduction to these techniques is tabulated in Table 3.1 for the basic understanding of readers. For detailed information, readers are encouraged to refer to process-specific references.

The above friction based processes can be further categorized into friction joining and friction processing techniques. Friction joining techniques can be subclassified as rotary friction based and non-rotary friction based processes. Rotary friction joining processes involve joining components with rotational symmetry, which includes processes like rotary friction welding, radial friction welding, and so on. Nonrotary friction joining processes include orbital friction welding, linear friction welding, and so on, where joining of noncircular components is involved. Friction processing techniques like friction extrusion, friction stir processing, and so on are utilized for recycling swarf and reprocessing and fabricating newer alloys, as well as composites.

3.5 Hybrid Friction Based Additive Manufacturing Processes

There are several friction based processes. Each of the processes has its own significance and limitations. But one thing common in all the processes is that they work in the solid state, that is, no melting. Owing to this, these processes result in fine-grained microstructures, low distortion, no liquid–solid phase transformation or related defects, the ability to produce wrought microstructures, and the ability to join dissimilar materials. These benefits make friction based processes suitable to couple with other manufacturing processes to produce high-performance end products. There are several friction based hybrid processes; however, according to the scope of this book, only those hybrid friction processes, which are based on additive manufacturing are discussed. These hybrid processes can be grouped as friction additive techniques. In all FATs, friction joining principles are coupled with the layered manufacturing approach of AM. These hybrid (in

TABLE 3.1

Friction Joining and Processing Techniques

Technique	Working Principle	Schematic
RFW	This is the most common FW. Two variants: continuous drive and stored energy. Machine weld range is quite broad. This involves welding of two axially aligned pieces, one of which is fixed, and the other undergoes rotational motion driven by the power source (a motor in the case of CDFW and a flywheel in the case of IFW). These parts are brought in contact under frictional force, and their rubbing actions amounts to frictional heating and material plasticization. Finally, an upsetting force is applied to consolidate the weld.	(a) CDFW (b) IFW, Flywheel
Linear and angular friction welding	In linear friction welding, one part traverses linearly and is pressurized against a stationary part to plasticize the material. Amplitude is then lessened to zero and forging force is applied to obtain frictional welding of the two parts. Angular friction welding works on a similar principle and can be extended to parts having angular alignment with respect to each other. Nonrounded intricate shaped parts displaying axisymmetry can be joined via this route.	
FS	This can be considered similar to friction deposition in that material addition takes place from a consumable rod in both cases. In FS, a consumable rod is utilized for depositing surface coating layers by metallurgical bonding via frictional heating. This technique is successful for substrates of any shape. Material addition from the consumable rod over the desired area takes place when the substrate is allowed to traverse.	Force, Consumable, Consumable flash, Deposit, Laterally moving plate

(*Continued*)

TABLE 3.1 (*Continued*)
Friction Joining and Processing Techniques

Technique	Working Principle	Schematic
Friction transformation hardening	This is a friction processing technique. Here, a rod that is not consumed during operation is rotated at preset peripheral speeds and imposed to touch and traverse along the substrate to be processed.	
Friction stir welding	This was developed at The Welding Institute in 1991 in U.K. It is one of the most important variants of FW and utilizes a nonconsumable rotating tool for the joining of similar and dissimilar materials. In this process, a shouldered pin tool is plunged into mating surfaces to be joined. Relative motion between tool and work is then accomplished. The rotational motion of the tool pin heats and plasticizes material to accomplish weld consolidation.	
Friction seam welding (Klopstock)	This is also utilized as a surfacing technique. A filler rod is employed to obtain the joint, and many metals and alloys can be friction welded using this technique. It can be utilized to obtain wear-resistant and anticorrosive coatings and also to repair damaged parts.	
Friction seam welding (LUC)	This is another example of friction seam welding where a nonconsumable round wheel rotates at extremely high peripheral speeds and simultaneously applies a force. Here, the wheel is treated as third body for obtaining frictional energy and plasticization. It is used to join metals having thinner gauges.	

(*Continued*)

Friction Based Joining Techniques 47

TABLE 3.1 (*Continued*)
Friction Joining and Processing Techniques

Technique	Working Principle	Schematic
Friction taper stitch welding	This involves machining of a tapered stud. This stud is then friction-welded into the hole/cavity, which is either machined or appears in the form of a crack. The process is repeatedly performed until there is overlapping of the welds. Tapered studs are required. This can accomplish repairs in hostile environments.	Crack
Radial FW	The ends of pipes/tubes to be joined are prepared in the form of a V shape. An internally bevel profiled ring is positioned in this V. Flash formation is prevented by providing an internal support. This ring is radially compressed for frictional heating and weld formation. It can be utilized to weld long pipes and tubes and can attach rings to cylindrical components.	Support mandrel / Stationary clamped pipes / Ring rotating and compressed
Orbital FW	This friction welding involves the initial alignment of the components about their common axes and their subsequent rotation in one direction at equal speeds to enable generation of orbital motion and obtain orbital amplitudes via application of frictional force. This sequence is halted by returning orbital amplitudes to zero and subsequently applying forging forces.	Orbit

(*Continued*)

TABLE 3.1 (*Continued*)
Friction Joining and Processing Techniques

Technique	Working Principle	Schematic
Friction hydropillar processing	This involves insertion of a rod undergoing rotation motion inside a hole/cavity and then advancing it toward the bottom under frictional force with an aim to obtain plasticization of the material and ensure that the material is transferred from the rod to fill the cavity. This is for repairing work in hostile environments, for example, under water, heavy radiations, dangerous explosion-prone environment, seaside, and aircraft applications. It is also a friction processing technique. The effectiveness of this technique for many copper-based alloys has been established, and as-cast microstructures can be overcome to obtain finely granular microstructures.	
Friction extrusion and friction co-extrusion cladding	This is also basically a surface processing technique. It can have different variants depending upon the function. It can be used to extrude a rod/billet to reduce cross-section or obtain wires. It can be used to clad a solid rod or a tube. It can also be used to consolidate powder and draw it in the form of wires. Different forms of dies are utilized to accomplish different functional requirements.	

(*Continued*)

TABLE 3.1 (*Continued*)
Friction Joining and Processing Techniques

Technique	Working Principle	Schematic
Friction stud welding	This friction welding method involves rotational movement of a solid stud, which is then forced into a plate surface in a monitored environment. The rotational motion of this stud during its downward motion causes plasticization at the interface, after which the motion is terminated and a forging force is applied for weld consolidation. This can be used for repairs in hostile environments.	Positioning of stud — Evaporation of ignition tip — Burning of arc — Stud in weld position (Basic weld unit, Ignition tip)
Friction brazing	This can be considered a special class of third-body friction welding, where a braze alloy layer is amassed upon the first part. Heating is accomplished by melting brazed alloy by closely monitoring speed and frictional force. Welding is then accomplished via brazing.	Braze alloy; Low axial force
Third body FW	In this friction welding method, a third body of relatively low melting point is utilized for metallurgical bonding of two primary parts. This basically involves placement of third-body material in granular form in the predrilled crevice (of a specified geometry) of one of the parts to be welded. The second part, with specific geometrical features, rotates and pushes forward to generate the heat of friction, causing plasticization of the third material, which consolidates the two parts to be welded.	Stud; Containment shoulder; Third body relatively low melting point material

(*Continued*)

TABLE 3.1 (Continued)
Friction Joining and Processing Techniques

Technique	Working Principle	Schematic
Friction plunge welding	This is a process to weld two components of variable hardness. Machining is carried out on the harder material of the pair to be welded to have locking features. It is then rotated against the softer one with the aim to achieve metallurgical bonding between the two materials.	*Without containment shoulder — Displaced relatively soft material; With containment shoulder — Relatively soft material being forced back onto the relatively hard material*
Friction deposition	This can be considered a modified version of RFW. In both cases, circular rod-shaped parts rub against each other. However, contrary to the joining of two parts as in the case of RFW, the material addition takes place from a consumable rod in the case of FD. When the rotating consumable rod is rubbed against the stationary rod, friction heat is generated and material addition takes place from the consumable rod onto the substrate on termination of the rotational motion.	*Onset — Deposit — Final; Consumable rod, Stationary substrate, Hot plastic material, Flash, Deposited material, Friction force*
Friction stir processing [25]	This is basically a derivative of FSW. Instead of welding or joining, as in the case of FSW, FSP is utilized for surface modification, composite/surface composite fabrication, alleviating defects, and recycling and reprocessing of materials and so on. However, the basic principle of both these processes is same.	*Shoulder, Pin, Process direction, Advancing side, Retreating side, Nugget*

(*Continued*)

TABLE 3.1 (Continued)
Friction Joining and Processing Techniques

Technique	Working Principle	Schematic
Friction stir additive manufacturing	FSAM utilizes the principle of layer-by-layer AM. Friction stir lap welding (FSLW) is generally utilized to additively join metals layer by layer. In its conventional form, a nonconsumable tool is inserted into the build/stack of overlapping sheets/plates and FSLW is carried out along the defined direction with optimum process parameters. It involves four basic steps of initial preparation, stacking, performing one FSLW, and then flattening upper surfaces. These steps are then repeated upto the desired build height.	
Additive friction stir	During AFS, material addition in the form of metal powder or a solid rod takes place from the center of a nonconsumable tool and flows in random directions owing to tool traverse motion. Tracks for each layer are overlapped, and each subsequent layer is deposited over the predeposited layer. After deposition of a layer over the substrate, the tool height is adjusted to accommodate for the deposition of the subsequent layer. Fillers may be induced in previous layers to achieve enhanced bonding. Due to frictional heating and plastic deformation owing to rotation of the tool w.r.t. the substrate, a strong metallurgical bond forms between successive layers.	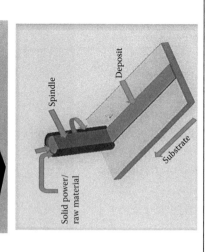

principle) techniques have mostly been developed in last decade and are proving themselves as suitable alternatives to fusion-based AM processes. These FATS are based on:

1. Rotary FW [26]
2. Linear FW [27]
3. Friction deposition [28]
4. Friction surfacing [29]

Based on friction stir welding:

5. Friction stir additive manufacturing [30]
6. Friction assisted (lap) seam welding [31]
7. Additive friction stir [32,33]

The progression of any process from its inception gives a general idea of how the process has evolved. This also helps to identify future trends, both in the vertical and horizontal space of further development. A timeline of the gradual development of these processes is presented in Table 3.2 and is discussed here: To eliminate the problems encountered in fusion-based AM techniques, in 2002, White [34] reported the consolidation of metallic materials using solid-state friction joining. Another report in this direction was presented by Lequeu et al. [35] in 2006 by fabricating Al 2050-Li wing ribs using friction stir welding-based additive manufacturing technology. After a due course of time, Dilip et al. [28] proposed two friction based AM techniques and suggested that friction welding and friction deposition can be successfully utilized in fabricating 3D components. After the reports of Baumann et al. [36] (Boeing) in 2012, the importance of FATs was recognized. They reported that the FSAM technique can be successfully utilized in fabrication of preform structures. In 2013, Kumar Kandasamy [37] filed a patent on AFS, and Dilip et al. [29] introduced a procedure of AM using FS. NASA completed a project on AFS during 2014–2016. Kalvala et al. [31] proposed friction-assisted (lap) seam welding as a novel procedure to develop 3D components. Palanivel et al. [38] accomplished fabrication of structural Mg-based WE 43 alloy through FSAM. They utilized FSAM to control microstructures of multilayered stacks of Mg-based WE43 alloy at two different welding parameters. Calvert [39] submitted a thesis on Aeroprobe's additive friction stir and develop/fabricate deposited magnesium alloy from powdered stock.

The list of research work included in Table 3.2 may not be exhaustive, but it is clearly evident from the table that, despite the early advent of FATs, limited research work has been accomplished in this innovative field, mainly owing to little information available in the literature.

Friction Based Joining Techniques

TABLE 3.2
Timeline of Friction Based Additive Manufacturing Processes

S. No.	Year	Remarkable Progress
1	2002	White [34] patented friction joining for AM.
2	2006	Lequeu et al. [35], Airbus, presented maiden report stating that FSW/P can be used for AM of metallic components.
3	2007	Threadgill and Russell [38] demonstrated rotary and linear friction welding as suitable options for MAM.
4	2011	Dilip et al. [28] introduced a novel friction based AM process and termed it friction deposition.
5	2012	Baumann et al. [36] (Boeing) reported that FSAM can achieve component fabrication via low production cost and lesser waste.
6	2012	Dilip et al. [26] reported that friction welding and friction deposition processes can be utilized as a probable AM route for fabricating high strength metallic components.
7	2013	Dilip et al. [29] demonstrated a probable method of friction surfacing for AM.
8	2013	Kandasamy et al. [39,40] proposed an additive friction stir process as a suitable route to fabricate defect-free aluminum and magnesium components.
9	2014	Kalvala et al. [41] patented friction-assisted seam welding as a feasible method of accomplishing AM.
10	2015	Palanivel et al. [30] used FSAM for microstructural control of magnesium-based alloys.
11	2015	Withers [42] suggested FSAM as a potential route for high-performing structural applications.
12	2015	Palanivel et al. [43] demonstrated FSAM as a route to fabricate high-performance aluminum- and magnesium-based alloys.
13	2014–2016	A NASA project was completed on additive friction stir technology that projected multiple advantages for the aircraft, space, and commercial industries, especially in terms of customization, achievement of wrought microstructures, low costs, etc.
14	2015	Calvert reported a thesis intended to study the microstructures and mechanical properties of components fabricated via AFS [44].
15	2016	Yuqing et al. [45] reported the fabrication of AA 7075 aluminium alloys build using FSAM.
16	2016	Kumar Kandasamy [37] granted patent on AFS.
17	2017	Palanivel and Mishra [27] reported a review work on friction based additive manufacturing processes.
18	2017	Rivera et al. [46] reported the fabrication of Inconel 625 using AFS.

3.5.1 Benefits and Limitations of Friction Based Additive Techniques

FATs have numerous benefits compared to liquid-phase AM techniques. The most common advantages of the processes are described here:

- All processes are based on the solid state
- No solidification-related defects like porosity
- Low distortion
- No cast microstructures
- Lower power requirements
- Can easily fabricate multimaterial and functionally graded materials
- Higher lateral strength as compared to fusion-based AM
- Can be utilized for a wide range of materials

In addition to common benefits, the common limitation of FATs (except AFS) is machining of deposited layers for shaping the build and preparing it for the addition of new layers. Besides the common features, each process has specific advantages and limitations, and these are presented in Table 3.3.

TABLE 3.3

Specific Benefits and Limitations of FATs

FATs	Benefits	Limitations
Rotary FW [26]	Suitable for large variety of metal alloys, can produce large build volumes	Not suitable for noncircular objects, nonhomogeneous microstructures
Linear FW	Suitable for objects other than circular	Poor bonding at edges
FD [28,47]	High deposition rate, excellent tensile strength, no requirement of filler	Unbonded regions at boundaries, suitable for circular parts
FS [29]	High layer thickness, no requirements of filler	Reduced mass transfer efficiency, not suitable for parts having downward-facing features or overhangs, poor bonding at edges
FSAM [30,45]	Broadened alloy space, no requirements of filler, engineering microstructures possible	Requires special fixtures for high thicknesses, susceptible to defects like hooking
FASW [31,41]	Dissimilar material can be easily joined, higher strength	Dependent on machine variables, most suitable for thin sheets
AFS [37,44,46]	Purely additive in nature, both powders and metal rods can be used as initial material, good lateral strength	Complex design of tool, dependence on machine variables

3.6 Conclusions

It can be concluded that FATs can be successfully utilized as a suitable alternative to fusion-based MAM processes. The specific advantages of these techniques as compared to fusion-based AM are their considerably reduced energy consumption, optimal part consolidation, and structural efficacy. These have ability to fabricate larger size single as well as multimaterial components. Components (fabricated via FATs) possess greater reproducibility rates, excellent mechanical properties, tailor-made microstructures, and so on. However, major challenges that restrict complete commercialization of these techniques are high initial capital investment and the inability to fabricate extremely intricate structures.

References

1. Maalekian, M. Friction welding—Critical assessment of literature. *Science and Technology of Welding and Joining*, 2007. 12(8): 738–759.
2. Bevington. Welding, T.N. 1891. Google Patents.
3. Bishop, E. Friction welding in the Soviet Union. *Welding and Metal Fabrication*, 1960. 28(10): 408–410.
4. Richter, W. Herbeifuhrung einer Haftverbindung zwischen Plattchen aus Werkzeugstahl und deren Tragern nach Art einer SchweiBung oder Lotung, 1929. Patent Number DE 476 480.
5. Vill, V. I. *Friction Welding of Metals, V. I. Vill: Translated from Russian* 1962, American Welding Society: New York.
6. Kauzlarich, J. J., Maurya Ramamurat, R. Reciprocating friction bonding apparatus. 1969. Google Patents.
7. Olson, D. L., Siewert, T. A., Liu, S., Edwards, G. R., Handbook committee. *Welding, Brazing and Soldering, ASM Handbook* (ed. D. L. Olson et al.). Vol. 6. 1993. ASM International: Materials Park, OH. pp. 503–515, 879–886.
8. Thomas, W. M., Nicholas, E. D., Needham, J. C., Nurch, M. G., Temple-Smith, P., Dawes, C. Friction stir butt welding. G.B. 1991, USA. International Patent Appl. No. PCT/GB92/02203 and GB Patent Appl. No. 9125978.8 (1991). Issuing Organization: TWI. Patent No. 5,460,317.
9. Nicholas, E. D. Friction processing technologies. *Welding in the World*, 2003. 47(11): 2–9.
10. Wang, K. K. Friction welding 1975. Welding Research Council, ISBN: 9781581452037, https://books.google.co.in/books?id=T7FeGwAACAAJ.
11. Nicholas, E. D. Linear friction welding, in *DVS Berichte Conference on Flash Butt and Friction Welding with Allied Processes* December 1991, Stuttgart, 18–24.
12. Dobrik, A., Gellerman, M. M. Increasing the life of hot rolling rolls by friction hardening. *Materials Science*, 1983. 19(3): 248–249.

13. Klopstock, H., Neelands, A. R. An improved method of joining or welding metals. 1941. British Patent Specification, 572789.
14. Luc, P. J. V., *Bonding aluminium*, 1975, Google Patents, http://www.google.mg/patents/US3899377.
15. Andrews, R. E., Mitchell, J. S. Underwater repair by friction stitch welding. *Metals and Materials*, 1990. 6(12): 796–797.
16. Teng, J., Wang, D., Wang, Z., Zhang, X., Li, Y., Cao, J., Xu, W., Yang, F. Repair of arc welded DH36 joint by underwater friction stitch welding. *Materials and Design*, 2017. 118(Supplement C): 266–278.
17. Nicholas, E. D. Radial friction welding. *Welding Journal*, 1983. 63(7): 17–28.
18. Xiaoling, X., Wei, W., Yuanze, X., Aiming, D. The research of radial friction welding. *Welding in the World*, 2005. 49(1): 31–33.
19. Searle, J. Friction welding non-circular components using orbital motion. *Welding and Metal Fabrication*, 1971. 39(8): 294–297.
20. Nicholas, E. D. Friction hydro pillar processing in advances in welding technology, in *11th Annual North American Welding Research Conference 7–9 November 1995, Columbus, Ohio*.
21. Thomas, W. M., Nicholas, E. D. Friction extrusion technology, in *TWI, Connect* March 1992.
22. Dittman, B. Friction brazing—An alternative to friction welding. *Schweisstechnik (Berlin)*, 1976. 26(2): 80.
23. Thomas, W. M., Nicholas, E. D., Jones, S. B. Third-body friction joining, in *Connect* April 1994.
24. Wayne, T., Welding: Friction takes the plunge, *Assembly Automation*, 1994. 14(1): 15–16.
25. Rathee, S., Maheshwari, S., Siddiquee, A. N. Issues and strategies in composite fabrication via friction stir processing: A review. *Materials and Manufacturing Processes*, 2018. 33(3): 239–261.
26. Dilip, J. J. S., Janaki Ram, G. D., Stucker, B. E. Additive manufacturing with friction welding and friction deposition processes. *International Journal of Rapid Manufacturing*, 2012. 3(1): 56–69.
27. Palanivel, S., Mishra, R. S. Building without melting: A short review of friction-based additive manufacturing techniques. *International Journal of Additive and Subtractive Materials Manufacturing*, 2017. 1(1): 82–103.
28. Dilip, J. J. S., Kalid R. H., Janaki Ram, G. D. A new additive manufacturing process based on friction deposition. *Transactions of the Indian Institute of Metals*, 2011. 64(1): 27.
29. Dilip, J. J. S., Babu, S., Varadha Rajan, S., Rafi, K. H., Janaki Ram, G. D., Stucker, B. E. Use of friction surfacing for additive manufacturing. *Materials and Manufacturing Processes*, 2013. 28(2): 189–194.
30. Palanivel, S., Nelaturu, P., Glass, B., Mishra, R. S. Friction stir additive manufacturing for high structural performance through microstructural control in an Mg based WE43 alloy. *Materials and Design (1980–2015)*, 2015. 65: 934–952.
31. Kalvala, P. R., Akram, J., Misra, M. Friction assisted solid state lap seam welding and additive manufacturing method. *Defence Technology*, 2016. 12(1): 16–24.
32. Rodelas, J., Lippold, J. Characterization of engineered nickel-base alloy surface layers produced by additive friction stir processing. *Metallography, Microstructure, and Analysis*, 2013. 2(1): 1–12.

33. Su, J. Additive Friction Stir Deposition of Aluminum Alloys and Functionally Graded Structures, Phase I Project SBIR/STTR Programs | Space Technology Mission Directorate (STMD), 2013. US.
34. White, D. Object consolidation employing friction joining. 2002. Google Patents.
35. Lequeu, P. H., Muzzolini, R., Ehrstrom, J. C., Bron, F., Maziarz, R. High performance friction stir welded structures using advanced alloys, in *Aeromat Conference* 2006, Seattle, WA.
36. Baumann, J. A. *Technical Report on: Production of Energy Efficient Preform Structures* 2012, The Boeing Company: Huntington Beach, CA.
37. Kandasamy, K. Solid state joining using additive friction stir processing. 2016. Google Patents.
38. Threadgill, P. L., Russell, M. J. Friction welding of near net shape preforms in Ti-6Al-4 V, in *11th World Conference on Titanium (JIMIC-5)* 2007, Kyoto, Japan.
39. Kandasamy, K., Renaghan, L. E., Calvert, J. R., Schultz, J. P. Additive friction stir deposition of WE43 and AZ91 magnesium alloys: Microstructural and mechanical characterization, in *International Conference, Powder Metallurgy & Particulate Materials* 2013, Chicago, IL, Advances in Powder Metallurgy and Particulate Materials.
40. Kandasamy, K., Renaghan, L., Calvert, J., Creehan, K., Schultz, J. Solid-state additive manufacturing of aluminum and magnesium alloys, in *Materials Science & Technology Conference and Exhibition 2013: (MS&T'13)* 2013, Montreal, Quebec, Canada, Materials Science and Technology-Association for Iron & Steel Technology.
41. Kalvala, P. R., Akram, J., Tshibind, A. I., Jurovitzki, A. L., Misra, M., Sarma, B. Friction spot welding and friction seam welding. 2014. Google Patents.
42. James Withers, R. S. M. Friction Stir Additive Manufacturing as a potential route to achieve high performing structures, 2015, US DOE workshop on Advanced Methods for Manufacturing (AMM): University of North Texas.
43. Palanivel, S., Sidhar, H., Mishra, R. S. Friction stir additive manufacturing: Route to high structural performance. *JOM*, 2015. 67(3): 616–621.
44. Calvert, J. R. Microstructure and mechanical properties of WE43 alloy produced via additive friction stir technology, in *Materials Science and Engineering* 2015, Virginia Polytechnic Institute and State University: Virginia Tech.
45. Yuqing, M. L., Ke, C., Huang, F., Liu, Q. L. Formation characteristic, microstructure, and mechanical performances of aluminum-based components by friction stir additive manufacturing. *The International Journal of Advanced Manufacturing Technology*, 2016. 83(9): 1637–1647.
46. Rivera, O. G. et al. Microstructures and mechanical behavior of Inconel 625 fabricated by solid-state additive manufacturing. *Materials Science and Engineering: A*, 2017. 694(Supplement C): 1–9.
47. Dilip, J. J. S., Janaki Ram, G. D. Microstructure evolution in aluminum alloy AA 2014 during multi-layer friction deposition. *Materials Characterization*, 2013. 86(Supplement C): 146–151.

4

Friction Joining-Based Additive Manufacturing Techniques

4.1 Introduction

The use of welding techniques in additive manufacturing (AM) technology is continuously increasing [1]. With the numerous advantages of AM in terms of ease of fabrication of complex geometrical parts, reduced material wastage, and so on, it has become the topmost choice in design and fabrication of parts in sectors like aerospace, automobile, transportation, medicine, and so on. Many advances have been reported on AM of metallic components in the last two decades. The utilization of metal additive manufacturing (MAM) has extended from the laboratory scale to the industrial sector, and it is being adopted by key industries [2,3]. As AM techniques produce three-dimensional components by progressively adding thin layers of material, the addition of these layers can be performed using different welding processes. Out of several welding processes, the most common being used for MAM include arc welding (MIG, TIG, plasma), laser beam welding, electron beam welding, and ultrasonic welding [1,4]. Except ultrasonic welding, all of these techniques basically work in liquid phase (molten pool) which generally results in liquid-solid-phase transformation defects. However, the ultrasonic welding–based method (known as ultrasonic consolidation/ultrasonic additive manufacturing) is the first solid-state AM technique that overcomes solidification-related microstructural defects. However, it suffers from its own limitations, as described in Chapters 1 and 2. Recently, friction welding/processing-based AM techniques have come into play [5,6]. These processes also work in the solid state and eliminate numerous problems associated with fusion welding–based MAM techniques [6]. Currently, as many as seven friction based AM processes are widely used (as described in Chapter 3). These processes have been categorized into different classes based on the mechanism of layer-by-layer material deposition/addition. Out of these, one classification is based on friction welding techniques, that is, rotary FW and linear FW-based AM techniques. Friction welding-based AM techniques are gaining importance, especially in cases that use dissimilar material makeups, owing to the fact that these processes are considered the most suitable techniques for dissimilar material joining. In addition to dissimilar material joining, all of the forgeable materials (metals) can be joined using FW [7]. Further, recent and future trends suggest greater use of parts with better structural and mechanical efficiency, which necessitates the use of customized materials. Most conventional processes are not capable of producing parts with such diverse properties in a single component. Specialized joining processes such as FW-based processes are capable of imparting the requisite characteristics in the fabricated components. Such newer materials that are alloyed and heat-treated via a complex combination of processes are sensitive to the underlying phenomena involved in fusion welding. Therefore, there has been a tremendous increase in research in the domains of solid-state welding, especially FW. In general, the underlying physics of both processes, that is, RFW and LFW, is identical and is based on the formation of metallic-bonded weld regions due to the heat-deformities effect at the contact surfaces. The basic aspects of these processes and their utilization for MAM are discussed in detail in the subsequent sections.

4.2 Rotary Friction Welding

RFW was the first commercial FWT and is quite popular even today. It is one the most common forms of FWT. It was initially patented by Bevington in 1891 for welding ends of rods and wires [8]. It is generally utilized to permanently join axisymmetric components using frictional heat generated by virtue of rotational motion of components with regard to each other. It is based on the principle of heat generation by directly converting mechanical energy to thermal energy that is utilized as the sole source of heat for weld formation. The frictional heat varies with the distance from the rotational axis and exhibits its lowest value at the center and highest values at the circumference of the component. It generally results in nonuniform properties of the weld.

4.2.1 Working Principles of Rotary Friction Welding

In RFW, one of the parts is kept in a still position while the other part undergoes rotational motion. The fixed part is forced against the rotating component and held in contact under constant or gradually increasing pressure. This stage is continued until the mating surfaces reach welding temperature. For continuing intimate contact of heated interfaces, axial pressure is further utilized. At this stage, diffusion occurs at the interface and constitutes a metallurgical bond at the faying surfaces. Thus, an RFW cycle constitutes of: (a) a friction phase that involves heating and upsetting and (b) a forging phase that involves weld consolidation. The process schematic is shown in Figure 4.1(a). There are five main common factors influencing the quality of weld during RFW [9]. These can be listed as:

- Applied pressure
- Relative velocity of surfaces
- Temperature of surfaces
- Properties of bulk material
- Condition of surfaces

The first three parameters are controllable during RFW, and last two are related to materials to be joined. RFW has two variants based upon the method of conversion of rotational energy to frictional heat. These are:

a. *Direct drive or continuous drive rotary FW* is also commonly called friction welding and is normally abbreviated as CDFW. It emerged around the 1940s [10]. In it, there are two axially aligned circular bars, one of which is rotating at a constant speed and is powered by a motor. The other remains stationary. These two bars are brought into physical contact at a prespecified force and time. Rotation continues until the metal to be joined reaches the plastic state. After this stage, the rotational movement is stopped and the forging force is adjusted with an aim to strengthen the joint. Thus, the CDFW process mainly consists of three phases, that is, a friction phase, forging phase, and termination phase [11]. Figures 4.1a and b present a schematic of CDFW and its welding parameters as well as the forces involved [12].

b. *Stored energy FW*, which is also known as inertia FW or flywheel FW, emerged in the 1960s and is normally abbreviated as IFW [10]. Here, the spindle that holds a component undergoing rotational motion is attached to a flywheel that is accelerated by a motor until it reaches predefined speed levels. The flywheel is then disengaged and the stationary component brought in contact with the rotating component. The faying surfaces of both work pieces rub against each other under welding force. Sometimes, additional forging force is required for ensuring sound welds. Figure 4.1c shows the relationship between welding parameters as well as the forces involved in IFW.

The main difference between CDFW and IFW is frictional speed. In CDFW, the friction speed remains constant, while in IFW, the frictional speed continuously decreases.

62   *Friction Based Additive Manufacturing Technologies*

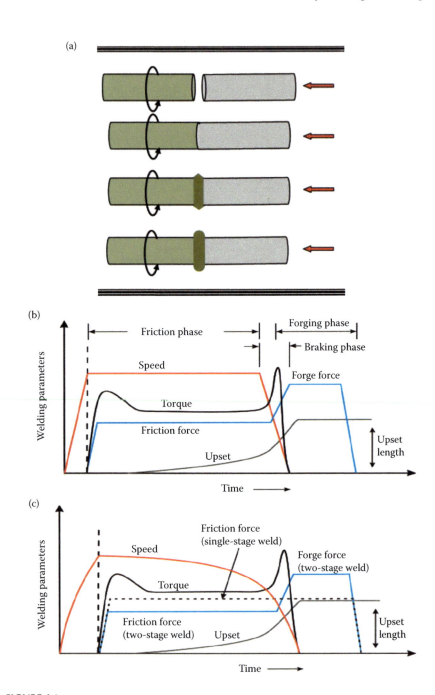

FIGURE 4.1
(a) Schematic of RFW variant. (Adapted from Maalekian, M. *Science and Technology of Welding and Joining*, 2007. 12(8): 738–759 [10]; Uday, M. B. et al. *Science and Technology of Welding and Joining*, 2010. 15(7): 534–558. [13]); welding parameters as well as forces involved in (b) CDFW, (c) LFW. (Adapted from Maalekian, M. *Science and Technology of Welding and Joining*, 2007. 12(8): 738–759.) [10]

4.2.2 Process Parameters Affecting Rotary Friction Welding

The process parameters affecting both RFW variants are basically identical. However, energy is supplied by different means in both, as discussed above. Therefore, the process parameters in both the cases are clearly enumerated below for clarity.

In CDFW, the following are the important process parameters:

a. Speed of rotation
b. Axial force
c. Time of frictional contact
d. Developed frictional pressure

In IFW, there are four main factors that need to be carefully controlled [10,11]. These are:

a. Moment of inertia for the flywheel
b. Speed of rotation
c. Axial force
d. Frictional pressure

In the CDFW process, axial force, speed of rotation, and welding time generally determine the rate of heat generation. Heat generation during various welding phases is variable [10]. During CDFW, axial shortening/burn-off/upset and resistive frictional torque vary during the process. The effect of process parameters and their relationship with CDFW are as shown in Figure 4.1b. Further, the relationship between speed of rotation and amount of generated heat can be numerically anticipated as: generated heat is proportional, while process time is inversely proportional to the speed of rotation [14]. This statement is in concurrence with the results of Li and Wang [15] during numerical analysis of CDFW of mild steel. However, Vill et al. [9] reported that at relatively high speeds, process time is directly proportional to speed of rotation, which can be attributed to the behavioral patterns of the material. Axial pressure has an effect on temperature gradients, welding power requirements, and axial shortening rates [14]. Corresponding to the materials being worked upon, the appropriate range of rotational speeds or axial pressures should be selected.

In IFW, the speed of rotation and axial pressure have almost same effects as those in CDFW. However, the rotational speed of the flywheel continuously decreases and depends on inertia and axial pressure. Wang et al. [16] suggested that increased axial pressure would make converting mechanical energies to heat easier in IFW. The kinetic energy of the flywheel has to be above a critical value, and selecting the right process parameters is critical for obtaining good welds. The relationships of process parameters and characteristics are shown in Figure 4.1c.

4.2.3 Additive Manufacturing with Rotary Friction Welding

The earliest demonstration of RFW as an AM technique was presented by TWI in 2007 [17]. Then, in 2012, Dilip et al. [5] demonstrated the procedure, benefits, and limitations of RFW as an AM tool to accomplish sequential addition of layers in a solid state. A sophisticated machine and setup has been utilized by Dilip et al. [5] to additively manufacture a remarkably large cross-section of 450 mm². Six layers were sequentially added to form a 78-mm (height) cylindrical object from AISI 310 stainless steel. It was reported that good bonding between different layers occurred, with fine-grained microstructures and good mechanical properties. Considerable grain refinement was reported near the near weld interface as compared to base metal. This grain refinement may be the result of the dynamic recrystallization that occurs during RFW. Surrounding the welded zone, thermomechanical-affected zones (TMAZs) as well as heat-affected zones (HAZs) appear that can be seen, as in the case of conventional friction welds.

Recrystallization is basically a restoration mechanism that leads to the formation of new grains free from dislocation or with lower dislocation densities [18]. Recrystallization occurring during plastic deformation at elevated temperatures is termed dynamic recrystallization (DRX). Metals with medium-stacking fault energy like (γ-Fe) austenitic stainless steel, nickel alloys, and so on generally undergo DRX. Copper (a medium-stacking fault energy metal) generally exhibits discontinuous DRX (DDRX). Metals with low-stacking fault energy frequently exhibit DDRX, while metals with high-stacking fault energy, like aluminum, exhibit continuous DRX (CDRX) when these materials undergo thermomechanical processing [19].

4.2.3.1 Applications of Rotary Friction Welding as an Additive Manufacturing Tool

A major benefit of utilizing RFW as an AM process is the freedom from component size restrictions and considerably reduced times. This can chiefly be attributed to the fact that the overall product volume depends on the constituent part dimensions, which in turn depend upon the machine capacities. Hollow (enclosed cavity) as well as multimaterial parts can be easily fabricated via this AM route.

A notable limitation of this process is the fact that it can be utilized only for fabrication of rounded parts showing axisymmetry like rods, shafts, and so on. Another limitation of this method is the coexistence of different zones, that is, nugget zone, TMAZ, HAZ, and base metal (BM), which are observed in a single layer leading to inhomogeneity in properties and the need of surface finishing owing to flash generation.

4.3 Linear Friction Welding

LFW is also called vibrational frictional welding and was first patented in 1929 [20,21]. However, active research in the field of LFW initiated during the 1980s, particularly in institutions like The Welding Institute in the United Kingdom. It is utilized to join parts displaying nonaxisymmetry with respect to one another.

4.3.1 Working Principles of Linear Friction Welding

Heat is generated by the action of frictional forces in low-frequency translation of one component with regard to others under pressure. Heat generated is more or less uniform, thereby amounting to higher uniformity in bonds as compared to RFW. This process basically consists of four distinct stages [22]. The first stage may be termed the initial stage, where abrasion of rough surfaces occurs at the interfacial regions when two objects with relative motion are brought together. Due to the abrasion of the asperities, the contact area starts increasing. The second stage may be termed the transition stage, where tattered surface features (that also contain impurities) are expelled and formation of fresh surfaces takes place, leading to achievement of the full contact area. Heating of the contact surfaces leads to softening of the material. This is an important stage that marks the building up of the temperature conditions necessary for successive stages. Determining an optimum condition requires a lot of experimentation, modeling, and analysis. It can be said in general that if there is uniform heating and material flow over contact surfaces, then welding circumstances are most favorable. The third stage may be termed the equilibrium stage, where an increase in axial pressure leads to extrusion and flash formation. During the third stage, the axial length is shortened and material is ejected from weld interfacial regions. When a predefined level of upsetting is achieved, parts are brought to rest at a rapid pace. Forging pressure can still be present for bond preservation. The fourth stage may be termed the deceleration stage, where consolidation of bonds takes place. It involves consolidation of bonds and affects the residual stresses created in bonds [23]. A schematic of LFW and these four stages are presented in Figure 4.2.

LFW utilizes very high-capacity, extremely sophisticated equipment as compared to RFW, and also demands close monitoring. This is especially important, as the vibrational motion originating due to oscillations of metallic masses necessitates careful dynamic balancing of systems and results in higher system cost. Specifications, size, constructional features, and details of the LFW

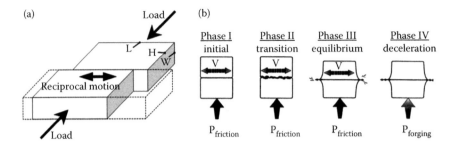

FIGURE 4.2
LFW process: (a) schematic, (b) four LFW phases. (Adapted from Vairis, A., Frost, M. *Materials Science and Engineering: A*, 1999. 271(1): 477–484; Li, W. et al. *International Materials Reviews*, 2016.) [24]

set up largely vary depending upon the needs of the part being joined [25]. The following are the main functional units that constitute a LFW setup:

- Generation unit to impart linear/angular oscillations to the movable element of the frictional pair
- Powering unit to impart frictional and upset force to the joint
- Fixing and positional units for correctly positioning welded elements
- Regulatory and control units to control parameters and regulate processes
- Shielding unit to specially weld active metals

4.3.2 Factors Affecting Linear Friction Welding

There are many important factors that affect the weld quality. A list of these is presented below:

1. Geometry of disc–blade contacting surfaces
2. Geometry of projections
3. Frequency as well as amplitude of oscillations
4. Intersection line of surfaces in contact
5. Method of generation of oscillations
6. Method to fix moving components
7. Method to protect edges from damage during oscillation
8. Profile of blade
9. Upsetting/burning off and its rate
10. Factors affecting heat generation
11. Frictional and forging pressures
12. Time of welding

Friction Joining-Based Additive Manufacturing Techniques

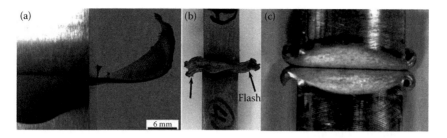

FIGURE 4.3
Images of LFW joints for various metals: (a) Ti-6.5Al-1.5Zr-3.5Mo-0.3Si. (Adapted from Lang, B. et al. *Journal of Materials Science*, 2010. 45(22): 6218–6224.) [28] (b) Carbon steel (medium). (Adapted from Li, W. Y., et al. *Materials Letters*, 2008. 62(2): 293–296.) [29] (c) AA 2024. (Adapted from Rotundo, F. et al. Linear friction welding of a 2024 Al alloy: Microstructural, tensile and fatigue properties, in *Light Metals 2012*, (ed C.E. Suarez), 2016, Springer International Publishing: Cham. pp. 493–496.) [30]

Since LFW, to a considerable extent, is specific to BLISK (blade + disk, a composite turbo engine part) fabrication, some of the above specific factors like blade profile have also been highlighted. The extent of axial shortening is treated as an indicator of sound joint fabrication by LFW by industries in general cases.

Some of the important observations reported by various researchers in the literature regarding the effect of important process parameters are discussed here. Wanjara and Jahazi [26] reported that weld time varies inversely with each process variable and power input. They further reported that there is threshold process condition to attain good-quality welds. Bhamji et al. [27] suggested that weld pressure varies directly with burnoff rates and inversely with friction time after threshold limits. Oscillation sense, material flow, and geometrical features are other important parameters with an effect on interfacial temperatures and weld shapes. The typical appearance of LFW joints and flash formation is shown in Figure 4.3.

4.3.3 Additive Manufacturing with Linear Friction Welding

AM using LFW was initially proposed by the Boeing Company [31]. Its researchers suggested an advancement of LFW where components of any shape and size can be fabricated by assembling basic defining shapes and then joining them using LFW. If one intends to forge a huge part like a titanium BLISK from a huge casting, the resulting process would be extremely expensive and complex. However, fabricating parts to approximate shapes using LFW to join smaller parts and intermediate machining is a far easier process. The savings in terms of material can be huge (of the order of 90%) as compared to conventional manufacturing of these parts [32].

The basic mechanism involved is quite simple. The first cycle of LFW comprising four phases (as described above) forms the initial substrate and consists of joining two initial smaller parts. The LFW flash is removed by

CNC machining. The next cycle is repeated by joining the third smaller part with the previously developed substrate, and the process is repeated until the desired final component is achieved [6]. The biggest advantage of using LFW is the fact that it can be extended to intricate geometries as compared to RFW, which is constrained to fabricate axisymmetric parts only.

4.3.3.1 Applications of Linear Friction Welding as an Additive Manufacturing Tool

The huge spectrum of applications offered by LFW can chiefly be attributed to the higher degree of parameter reproducibility and excellent weld quality. It is used for welding many critical joints in a given structure, which can in turn have varied applications. Several giants like Rolls-Royce, United Technologies, the Welding Institute, MTU Motoren and Turbinen, Boeing, and so on have an appreciable number of patents in LFW and are pursuing a lot of active research in this field. While the former three have utilized it for constructing engine parts, LFW has also been used for aircraft, transport, wheel rims, braking blocks, complicated transport parts, and so on by the latter two. Compressor designs of aircraft engines as well as gas turbines are high-end technological components. Their designs are extremely critical for performance of aircrafts and turbines. Thus, there is an ever-increasing need to improve their technical as well as environmental performance. Conventionally, these utilize a compressor disk in blade form with a discrete airfoil secured in place by nuts and bolts in the main retaining unit with slots. Recently, these have been merged into a single welded component called BLISK. An illustration is presented in Figure 4.4. These are also called integrated bladed rotors (IBRs) [12]. BLISK markets reflect a considerably increasing trend over the last 15 years, along with an increased present and forecast demand for fabrication and repair of BLISKs [33]. LFW technology is the world's leading technology dominating the BLISK manufacturing industry, especially in moderate- to huge-sized BLISKs. There are many underlying advantages of utilizing LFW for manufacturing BLISKs. A few of them are listed below:

- Elimination of mechanical joints, leading to appreciably increased tool life and weight reduction
- Elimination of stamping process for fabricating larger-diameter discs
- Minimal finishing requirements
- Innovative technology for repair of faulty and damaged blades, thereby appreciably improving engine working span
- Reduction in engine fabrication cost

Despite numerous applications and the versatility of LFW, there are many roadblocks that need to be dealt with before its large-scale commercialization.

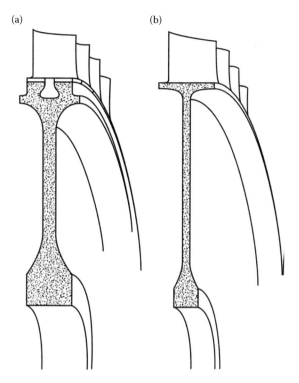

FIGURE 4.4
(a) Conventional bladed disc, (b) BLISK. (Adapted from Turner, R. et al. *Acta Materialia*, 2011. 59(10): 3792–3803.) [34]

One important issue that needs to be addressed is the availability of simpler setups for performing LFW. Another issue is related to the high cost of the setup. Other major limitation is the need to machine each layer, which adds a lot of expensive machining and material cost.

4.4 Comparison of Rotary Friction Welding and Linear Friction Welding

As evident from the detailed discussions on RFW and LFW in the above sections, there are many similarities and dissimilarities between them. These are presented in Table 4.1.

A huge spectrum of materials has been used in LFW and RFW. Figure 4.5 shows a comparison of various materials upon which research has been carried out for RFW and LFW.

TABLE 4.1

Comparison of RFW and LFW

S. No.	Rotary Friction Welding	Linear Friction Welding
Similarities		
1	Both are based on friction welding.	
2	Both are used for permanent joining of components.	
3	During the entire process in both cases, the material is in a solid state.	
4	The physics of both the processes is same and involves formation of metallic bonded welded regions due to the heat-deformities effect.	
5	Both are frequent techniques of welding dissimilar metallic parts.	
6	In both, tensile properties of fabricated parts are overmatched with base metals.	
7	In both, shield gases or filling materials are not required.	
Dissimilarities		
1	Most common FWT.	Relatively newer FWT.
2	Used to join axisymmetric components.	Used to join nonaxisymmetric components.
3	Weld has less uniform structure.	Weld has more uniform structure.
4	Requires relatively simple setup.	Requires quite expensive and complicated setup.
5	Maximum temperatures away from center, so HAZ is wider than in LFW.	Maximum temperatures obtained at centers, so HAZ of lower width than in RFW.

4.5 Advantages and Limitations of Friction Welding

Following are the advantages of LFW and RFW due to their solid-state nature:

1. Materials to be welded:
   a. Dissimilar material welding
   b. Welding materials impossible to join by fusion welding
   c. Can be utilized for special-purpose materials
   d. Can be utilized for functionally graded materials
2. Microstructural properties
   a. Fine-grained structures
   b. Optimal density of materials in welded joint
   c. High impact toughness
   d. Enhanced fatigue characteristics

Friction Joining-Based Additive Manufacturing Techniques

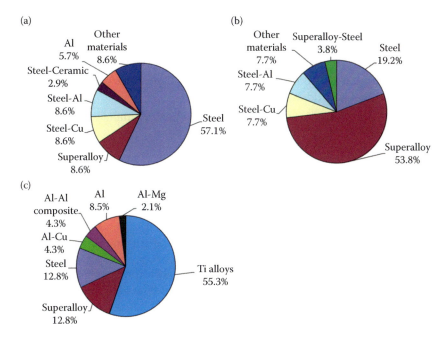

FIGURE 4.5
Proportion of materials upon which research has been reported for: (a) CDFW, (b) IFW, (c) LFW. (Adapted from Li, W. et al. *International Materials Reviews*, 2016. 61(2): 71–100.) [14]

3. Accuracy
    a. Highly accurate jointed structures
    b. High-quality welded joints
    c. Highly reliable
4. Apparatus required
    a. High automation in process
    b. Welding tools not required
5. Self-cleaning weld fabrication
6. Fast weld cycle
7. Almost complete area weld coverage

In addition to above listed benefits, FW suffers some limitations. Following are the limitations of LFW and RFW:

1. Require specialized equipment where energy needs to be automatically controlled
2. Process time parameters need to be closely controlled and monitored

3. Constrained ability of RFW to join nonaxisymmetric parts
4. Constrained ability of CDFW to weld bigger cross-sectional areas
5. Constrained ability to weld thin-walled components
6. Welding can only be accomplished when at least one component can undergo plastic deformation
7. Applications are limited to cases where part geometry allows flash removal

4.6 Conclusion

Based on experimental results and extensive literature surveys for RFW and LFW, it can be concluded that both of these techniques can be successfully utilized as additive manufacturing tools for the fabrication of 3D objects based on the friction welding route. They can be utilized on wide range of materials and are especially suitable for joining dissimilar metals owing to their solid-state nature. The occurrence of severe plastic deformation and subsequent dynamic recrystallization make these processes suitable for producing high-strength structural materials. RFW can be used in joining axisymmetric components. LFW can especially be used to join nonaxisymmetric components. The application of RFW in fabricating 3D objects with enclosed cavities is quite impressive. However, limited work has been done in this field; in particular, LFW has been utilized to additively manufacture and repair aircraft BLISKs only owing to the high cost and sophistication of the machine setup required. Also, research to overcome the geometric constraints of the parts to be joined will go a long way in improving the versatility of these processes.

References

1. DebRoy, T., Wei, H. L., Zuback, J. S., Mukherjee, T., Elmer, J. W., Milewski, J. O., Beese, A. M., Wilson-Heid, A., De, A., Zhang, W. Additive manufacturing of metallic components—Process, structure and properties. *Progress in Materials Science*, 2018. 92(Supplement C): 112–224.
2. Milewski, J. O. Additive manufacturing of metals. *Springer Series in Materials Science*. 258. 2017: 1–339, Springer.
3. Herzog, D., Seyda, V., Wycisk, E., Emmelmann, C. Additive manufacturing of metals. *Acta Materialia*, 2016. 117(Supplement C): 371–392.
4. Ding, D., Pan, Z., Cuiuri, D., Li, H. Wire-feed additive manufacturing of metal components: Technologies, developments and future interests. *The International Journal of Advanced Manufacturing Technology*, 2015. 81(1): 465–481.

5. Dilip, J. J. S., Janaki Ram, G. D., Stucker, B. E. Additive manufacturing with friction welding and friction deposition processes. *International Journal of Rapid Manufacturing*, 2012. 3(1): 56–69.
6. Palanivel, S., Mishra, R. S. Building without melting: A short review of friction-based additive manufacturing techniques. *International Journal of Additive and Subtractive Materials Manufacturing*, 2017. 1(1): 82–103.
7. O'Brien, R. L. *Welding Handbook*. 8th ed. Welding Processes. Vol. 2. 1991, American Welding Society: Miami.
8. Bevington, Welding, T.N. 1891. Google Patents, Publication number US463134 A, https://www.google.co.in/patents/US463134.
9. Vill, V. I. *Friction Welding of Metals*, V. I. Vill: translated from Russian 1962, American Welding Society: New York.
10. Maalekian, M. Friction welding—Critical assessment of literature. *Science and Technology of Welding and Joining*, 2007. 12(8): 738–759.
11. Olson, D. L., Siewert, T. A., Liu, S., Edwards, G. R. Handbook committee, *Welding, brazing and soldering, ASM handbook*, (ed D. L. Olson et al.). Vol. 6. 1993. ASM International: Materials Park, OH, pp. 503–515, 879–886.
12. García, A. M. M. BLISK fabrication by linear friction welding, *Advances in Gas Turbine Technology* (ed. E. Benini), 2011, InTech: Rijeka, Croatia.
13. Uday, M. B., Ahmad Fauzi, M. N., Zuhailawati, H., Ismail, A. B. Advances in friction welding process: A review. *Science and Technology of Welding and Joining*, 2010. 15(7): 534–558.
14. Li, W., Vairis, A., Preuss, M., Ma, T. Linear and rotary friction welding review. *International Materials Reviews*, 2016. 61(2): 71–100.
15. Li, W., Wang, F. Modeling of continuous drive friction welding of mild steel. *Materials Science and Engineering: A*, 2011. 528(18): 5921–5926.
16. Wang, F. F., Li, W. Y., Li, J. L., Vairis, A. Process parameter analysis of inertia friction welding nickel-based superalloy. *The International Journal of Advanced Manufacturing Technology*, 2014. 71(9): 1909–1918.
17. Threadgill, P. L., Russell, M. J. Friction welding of near net shape preforms in Ti-6Al-4V, in *11th World Conference on Titanium (JIMIC-5)* 2007, Kyoto, Japan.
18. Kalvala, P.R., Akram, J., Misra, M., Friction assisted solid state lap seam welding and additive manufacturing method. *Defence Technology*, 2016. 12(1): 16–24.
19. Huang, K., Logé, R. E. A review of dynamic recrystallization phenomena in metallic materials. *Materials & Design*, 2016. 111(Supplement C): 548–574.
20. Richter, W. Herbeifuhrung einer Haftverbindung zwischen Plattchen aus Werkzeugstahl und deren Tragern nach Art einer SchweiBung oder Lotung, 31 May 1929.
21. Nunn, M. E. Aero engine improvements through linear friction welding, in *1st Int. Conf. on 'Innovation and Integration in Aerospace Sciences'*. 2005, Belfast, UK.
22. Vairis, A., Frost, M. On the extrusion stage of linear friction welding of Ti 6Al 4V. *Materials Science and Engineering: A*, 1999. 271(1): 477–484.
23. McAndrew, A. R., Colegrove, P. A., Bühr, C., Flipo, B. C. D., Vairis, A. A literature review of Ti-6Al-4V linear friction welding. *Progress in Materials Science*, 2018. 92(Supplement C): 225–257.
24. Li, W. Achilles, V. Preuss, M. Ma, T. Linear and rotary friction welding review. *International Materials Reviews*, 2016 61(2): 71–100.
25. Vairis, A., Frost, M. Design and commissioning of a friction welding machine. *Materials and Manufacturing Processes*, 2006. 21(8): 766–773.

26. Wanjara, P., Jahazi, M. Linear friction welding of Ti-6Al-4V: Processing, microstructure, and mechanical-property inter-relationships. *Metallurgical and Materials Transactions A*, 2005. 36(8): 2149–2164.
27. Bhamji, I., Preuss, M., Threadgill, P. L., Addison, A. C. Solid state joining of metals by linear friction welding: A literature review. *Materials Science and Technology*, 2011. 27(1): 2–12.
28. Lang, B., Zhang, T. C., Li, X. H., Guo, D. L. Microstructural evolution of a TC11 titanium alloy during linear friction welding. *Journal of Materials Science*, 2010. 45(22): 6218–6224.
29. Li, W. Y., Ma, T. J., Yang, S. Q., Xu, Q. Z., Zhang, Y., Li, J. L., Liao, H. L. Effect of friction time on flash shape and axial shortening of linear friction welded 45 steel. *Materials Letters*, 2008. 62(2): 293–296.
30. Rotundo, F., Morri, A., Ceschini, L. Linear friction welding of a 2024 Al alloy: Microstructural, tensile and fatigue properties, in *Light Metals 2012*, (ed C. E. Suarez), 2016, Springer International Publishing: Cham. pp. 493–496.
31. Slattery, K. T., Young, K. A. Structural assemblies and preforms therefor formed by friction welding, 2008, Google Patents.
32. Addison, A. C. *Linear friction welding information for production engineering*, in TWI Industrial Member Report Summary 961/20102010: TWI, Granta Park, UK, 2010.
33. Shtrikman, M. M. Linear friction welding. *Welding International*, 2010. 24(7): 563–569.
34. Turner, R., Gebelin, J. C., Ward, R. M., Reed, R. C. Linear friction welding of Ti–6Al–4V: Modelling and validation. *Acta Materialia*, 2011. 59(10): 3792–3803.

5
Friction Deposition-Based Additive Manufacturing Techniques

5.1 Introduction

In this chapter, two deposition-based solid-state FATs are described and their salient features are discussed. The idea behind these processes comes from friction welding in which two mating surfaces rub together, resulting in generation of frictional heat and plastic deformation. Under these conditions, when axial force is applied, joining of the deformed surfaces takes place. Using this principle, multiple layers can be joined via friction welding, and by shaping/machining (using CNC), these layers can be given desirable shapes to form 3D objects. Using this technique, two processes, namely friction deposition and friction surfacing AM, were developed. These processes basically involve deposition of a consumable metal rod over a substrate. Both these processes are similar in principle, as, in both processes, the addition of material takes place from a consumable rod. In friction deposition, the addition of layers takes place at once for circular objects, while in the latter case, the addition of material takes place when the substrate (a noncircular object) traverses in a predefined direction.

5.2 Friction Deposition

FD is one of the simplest friction based processes that have been utilized for AM of metallic materials. It was introduced by Dilip et al. [1] in 2011 as a suitable technique for fabricating 3D metallic parts and is considered one of the most common friction based metal additive manufacturing processes.

In its basic operation, it involves a rotating consumable rod of defined dimensions being allowed to rub against a stationary substrate rod. These rods are fixed in axial alignment in machine spindles under axial load. The consumable rod is fixed in a rotating spindle, and the substrate (on which

FIGURE 5.1
Friction-deposited rough and machined surfaces of AISI 304 deposit. (Adapted from Dilip, J. J. S. et al. *Transactions of the Indian Institute of Metals*, 2011. 64(1): 27.) [1]

material is deposited) is held in a stationary spindle. When these two rods are brought together, frictional heat generates owing to the rubbing of metallic surfaces, resulting in plasticization of the material of the abutting surfaces. The development of increased quantities of plasticized materials occurs with the ongoing process, and the rotation of the rod is terminated by withdrawing it as soon as a suitable thickness of plasticized material is developed. A thickness of deposited material remains on the substrate after withdrawl of the consumable rod. This deposition of material on the substrate can be attributed to the detachment of plastified material from the distant end of the consumable rod owing to torsional shear stresses upon stopping the rotation and its withdrawal [1]. For depositing successive layers, initially machining of previously fabricated layers/build is done commonly by CNC, as friction deposits leave concave surface depressions after consumable rod detachment, as illustrated in Figure 5.1. Machining of the rough surface makes the ends flat and prepares it for the addition of subsequent layers. Thus, upon building/joining successive layers one over the other and subsequent machining, 3D parts can be fabricated. The common steps for AM of parts using FD are illustrated in Figure 5.2.

5.2.1 General Features and Experimental Results on Additive Manufacturing Using Friction Deposition

The process of FD as an additive manufacturing tool has recently been introduced and is attracting attention in the solid-state metal additive manufacturing sector. The efficiency of FD process and the properties of deposited material depend on the combination of process parameters utilized during the process. Major process parameters involve rotational speed of the consumable rod, applied axial force, diameter of the consumable rod, and so

Friction Deposition-Based Additive Manufacturing Techniques

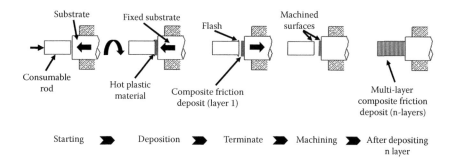

FIGURE 5.2
Steps in AM of 3D components via FD.

on. The rotational speed in almost all the friction based processes significantly affects heat generation. A high rotational speed implies more heat input and ultimately determines the bonding quality. Axial force in FD defines the consolidation of material. High axial force increases adhesion of the deposit while also decreasing the thickness of the deposit. However, with an increase in axial force, flash increases, which decreases process efficiency. The diameter of the consumable rod is highly interrelated to other process parameters and vice versa. A higher rod diameter increases the width of the deposit but demands high-capacity equipment. Its value should be carefully selected depending upon the type of material to be deposited and past experience with the process. Further, the optimum parameter combination depends on the type of materials under investigation. The process parameters selected (by various authors in the literature) during additive manufacturing (using FD) of different materials and their property enhancements are listed in Table 5.1.

In a nutshell, FD generally works on temperatures above recrystallization temperature and below the melting points of the materials being deposited. At such temperatures and under high strain rates, deposits produced via FD undergo dynamic recrystallization, which results in fine-grained microstructure; one such example is shown in Figure 5.3 [1]. Figure 5.3a shows an optical micrograph of as-received AISI 304 steel having an average grain size of 80 μm. After the FD process (process parameters and other details of this study are described in Table 5.1), the deposit/coating shows a refined grain structure, and equiaxed grains of average sizes of ~20 μm were achieved (refer to Figure 5.3b). This grain refinement further helps in enhancement of mechanical properties.

Little work is accomplished in the direction of AM using this process. However, FD as an additive manufacturing technique is successfully utilized for ferrous as well as nonferrous metal alloys by various researchers, as described in Table 5.1. A brief summary of general observations during experimental studies reported by various researchers on ferrous and nonferrous metal alloys is presented here for more clarity on the development and current status of the process.

TABLE 5.1
Process Parameters and Other Details Used in Some Experimental Studies on Friction Deposition Additive Manufacturing for Different Materials

Consumable Rod Material, Diameter (mm)	Substrate Material, Diameter (mm)	Process Parameters, w (rpm), p (kN)	No. of Layers	Dimensions of Deposit, H (mm), D (mm)	Property Enhancement	References
AISI 304 steel rod, 19 mm	Mild steel, 25 mm	w: 800 p: 9.8	30	H: 50 D: 20	YS: 380 (O)/YS: 390 (FD) UTS: 710(O)/UTS: 690(FD)	[1]
Borated stainless steel, 10 mm	AISI 304, 15 mm	w: 800 p: 8	More than 70	H: 40 D: 10	YS: 250 (O)/YS: 330 (FD) UTS: 620(O)/UTS: 700(FD)	[2]
Inconel 718, 10 mm	Inconel 718, 15 mm	w: 800 p: 8	More than 40	H: 40 D: 10	YS: 1150(O)/YS: 1200 (FD) UTS: 1410(O)/UTS: 1440 (FD)	[10]
AA5083-H112, 20 mm, filled with titanium powder	AA5083:H112, 25 mm	w: 800 p: 7	—	H: 40 D: 20	YS: 180 ± 4 (O)/YS: 210 ± 10 (FD) UTS: 300 ± 6 (O)/UTS: 350 ± 10 (FD)	[3]
AA5083-H112, 20 mm, filled with CoCrFeNi HEA powder	AA5083:H112, 25 mm	w: 800 p: 8	—	H: 40 D: 20	YS: 180 ± 4 (O)/YS: 280 ± 5 (FD) UTS: 300 ± 3 (O)/UTS: 395 ± 7 (FD)	[8]

Abbreviations: w: Rotational speed; p: force applied; H: height; D: diameter of rod; YS: yield strength; UTS: ultimate tensile strength.

Friction Deposition-Based Additive Manufacturing Techniques

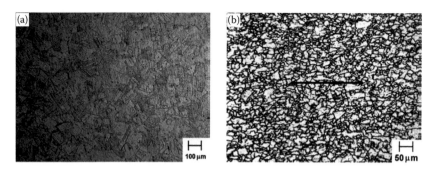

FIGURE 5.3
Optical micrographs of: (a) as-received base metal AISI 304 steel, (b) after FD. (Adapted from Dilip, J. J. S. et al. *Transactions of the Indian Institute of Metals*, 2011. 64(1): 27.) [1]

5.2.1.1 Development of Ferrous Metal Deposits Using Friction Deposition

5.2.1.1.1 AISI 304 Steel

Dilip et al. [1] proposed and reported maiden results on 3D fabrication of AISI 304 steel using FD. In this work friction welding machine was utilized to fabricate 50 mm thick deposit on mild steel substrate. This was accomplished in layered fashion by depositing more than 30 layers. Uniform thickness of layers was achieved with higher yield strength of deposit as compared to base metal (BM). The higher yield properties were achieved owing to fine recrystallized grained microstructures produced by FD. The details of enhancement of properties are listed in Table 5.1.

5.2.1.1.2 Borated Stainless Steel

FD of about 70 layers of borated stainless steel on austenitic stainless steel AISI 304 was performed by Dilip et al. [2]. Figure 5.4a shows a longitudinal section of the friction deposit. After microstructural examination, it was found that all the layers were well bonded to each other at their interfaces and no unbonded region or defect occurred at the interfaces, as shown in Figure 5.4b. However, at the outer periphery of each layer, some unbonded regions were present (refer to Figure 5.4c), which did not demand much attention, as they were machined during the shaping of layers by CNC machining [2].

In addition to good bonding between layers, FD produces fine-grained microstructures (as already discussed) owing to severe plastic deformation and subsequent recrystallization. The grain size of BM alloy 304B4 Grade B (40 μm) reduces drastically during FD coating (~2 μm) without formation of deleterious phases. An interesting observation in the case of FD of borated steel is that this material has no negative effect of a repeated thermal cycle, as generally occurs in heat-treatable alloys [2]. This is owing to the fact that the peak process temperature during FD process was measured as 1130°C, which is appreciably less than the eutectic melting point of borides in alloy 304B4. And, according to a study conducted on heat exposure of this alloy, it was

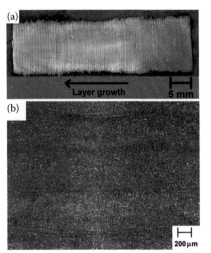
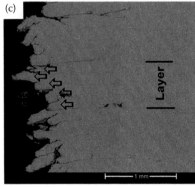

FIGURE 5.4
(a) Macrographic image of longitudinal section of deposit produced by FD, (b) optical microstructural images of deposit, (c) SEM image taken from edge of deposit. (Adapted from Dilip, J. J. S. et al. *Journal of Materials Engineering and Performance*, 2013. 22(10): 3034–3042.) [2]

found that no coarsening of boride particles occurs even at 1175°C for 4 hours of heat exposure [2]. Thus, from these results, it can be concluded that FD has high suitability in producing deposited 3D components of borated steels.

5.2.1.2 Development of Nonferrous Metal Alloy Builds Using Friction Deposition

It is generally assumed that nonferrous alloys of metals like aluminum, magnesium, and so on are affected due to thermal exposure. At elevated temperatures, these alloys experience an annealing effect and deleterious phase formation occurs, which results in coarsening of grains and decreasing of mechanical properties. The general features of the development of 3D parts of nonferrous metal alloys using FD are discussed here.

5.2.1.2.1 Development of Metal–Metal Composites via Friction Deposition

Metal–metal composites have gained attraction in the last decade. In such composites, metal particles are used as reinforcement to embed in a metal matrix substrate. These composites eliminate the problems of lower ductility and poor bonding between reinforcement and substrate (as in the case of metal–ceramic composites) to a large extent. The problem of poor bonding between ceramic particles and metal may occur owing to large variations in the coefficient of thermal expansion and elastic modulus of the ceramic reinforcements and metal matrix. Metal–metal composites exhibit a good combination of strength and ductility [3]. Several processes like powder metallurgy, disintegrated melt deposition, friction surfacing, and friction stir processing have been successfully utilized in developing metal–metal composites [4–7]. However, fabrication of these composites using the conventional AM processes has not been reported until now. MAM techniques like LENS and SLM have not been utilized for such composite fabrication owing to various reasons stated in Chapter 2. Binder printing techniques are currently utilized to fabricate these composites, but these processes involve postprocessing at elevated temperatures, which further results in deterioration of mechanical properties. Based on these facts, it can be concluded that there is a strong need to develop solid-state MAM techniques for circular objects that can overcome these problems. Recently, FD has been proposed for fabrication of metal–metal composites in layer-by-layer fashion [3]. In the subsequent paragraphs, fabrication of these composites using an FD-based AM method is discussed.

Using, FD, AA 5083-H112-based metal–metal composites were fabricated [3,8]. In these works, AA 5083-H112 was utilized as a consumable rod as well as substrate material. To fabricate composite deposits, titanium powder and nanocrystalline CoCrFeNi high-entropy alloy powder were initially filled into holes drilled in the consumable rod, then FD was performed. The friction deposits (longitudinal direction) are shown in Figure 5.5. From

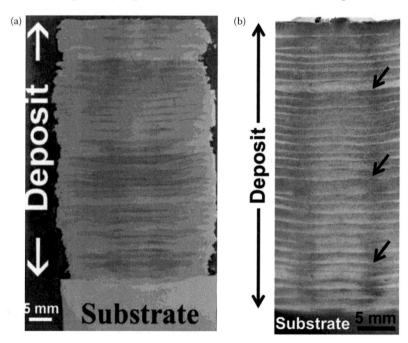

FIGURE 5.5
Macrographic image (longitudinal section) of friction deposit of: (a) AA 5083-titanium composite. (Adapted from Karthik, G. M. et al. *Materials Science and Engineering: A*, 2016. 653(Supplement C): 71–83.) [3]; (b) AA5083-nanocrystalline CoCrFeNi composite. (Adapted from Karthik, G. M. et al. *Materials Science and Engineering: A*, 2017. 679(Supplement C): 193–203.) [8]

82 Friction Based Additive Manufacturing Technologies

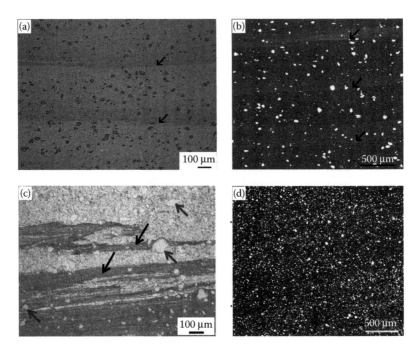

FIGURE 5.6
Microstructural images of composite region showing distribution of reinforcement particles: (a) optical micrograph of AA5083-titanuim, (b) SEM micrograph of AA5083-titanuim. (Adapted from Karthik, G. M. et al. *Materials Science and Engineering: A*, 2016. 653(Supplement C): 71–83. [3]) Arrows in the figure show interfaces of different layers; (c) and (d) are optical and SEM micrographs of AA5083-CoCrFeNi composite. (Adapted from Karthik, G. M. et al. *Materials Science and Engineering: A*, 2017. 679(Supplement C): 193–203. [8]) Black arrows show deformation shear bands, while red arrows show reinforcement particles.

microstructural images (refer Figure 5.6), it can be observed that FD produces defect-free, homogeneously distributed composites. Good bonding between successive layers takes place without formation of any intermetallics. The severe plastic deformation and subsequent dynamic crystallization results in enhanced mechanical properties. In addition to this, uniform dispersion of reinforcement particles and the pinning effect of reinforcements add to the mechanical strength of the resulting composites [9]. The properties of the fabricated composite build are listed in Table 5.1.

5.2.2 Benefits and Limitations of Friction Deposition

AM of metallic components using the FD method has the following advantages over fusion-based MAM processes [1,11]:

1. Layer-by-layer fabrication of material takes place in solid state, possessing excellent bonding strength among layers as between consumable and substrate.

2. Higher deposited layer thickness (achieved up to 1 mm), which is comparatively larger than commercially available MAM techniques.
3. One complete layer can be produced in a single run in relatively lesser time.
4. Produces fine-grained microstructure comparable to wrought microstructures.
5. Parts fabricated via FD exhibit excellent tensile strength.
6. Functionally graded materials and multimaterial components can be easily fabricated.
7. Homogeneously distributed composites can be successfully fabricated.

Apart from the above-listed benefits, the FD-based AM method suffers from very few limitations, such as the machining of each layer for shaping it to the desired shape, as poor bonding takes place at the edges of the deposits.

5.3 Friction Surfacing

Friction surfacing is a well-accepted, prominent technology for material deposition in the form of coatings in the solid state. The idea of FS was initially patented in 1941 [12]. However, similar to other novel ideas, the work on FS was not very popular for a considerable span after its advent. A few reports summarized by Bishop [13] indicated that this process was also developed during the 1950s in the USSR. However, research addressing FS stayed relatively dormant during the following decades. Even to date, the process has been chiefly limited to the laboratory scale, and its full-scale utilization is limited to stationary applications in flat joining positions. The chief factor behind this can be the high process forces requiring the use of large and rigid machine tools. Since the late 1980s, this process has gained a lot of attention following the growing interest in friction based solid-state processes [14]. In the search for superior coating solutions, FS is being increasingly investigated to obtain homogeneous fine-grained coatings that show better wear and corrosion resistance [15].

In FS, layer-by-layer addition of material can be done in which subsequent layer addition takes place via solid-state welding without fusion. FS is also a variant of friction welding [14]. FS is the solution to many problems encountered during the production of metallic coatings. It is similar to FD in principle but differs in its working, since in FS, deposit takes place on a noncircular substrate when it travels in a predefined direction, while in FD, no movement of the substrate takes place. Despite its early development in the 1940s, its usage as an AM tool has been recently proposed by Dilip et al. [16].

5.3.1 Working Principles of Friction Surfacing

FS utilizes a rotating consumable rod (which will be referred to as the rod or mechtrode hereafter) that is pressed against a substrate under axial pressure. Heat generation takes place owing to friction between the mating surfaces of the rod and substrate metal, resulting in softening and plasticization of the rod tip material. After adequate plastification (initial), the substrate is allowed to traverse in a predefined direction. With the continuous feed of the substrate, the plastified material from the rod tip continuously deposits onto the substrate. The schematic of FS is shown in Figure 5.7. The thickness and width of the deposited material depends on the types of material being processed and the selected process parameters. Generally, the width of the deposited material is the same as the diameter of the consumable rod [16].

Although FS has been utilized a long time as a surface improvement or metal deposition technique, its utilization for additive manufacturing is quite recent and is quite interesting. FS can be utilized as an AM process in the following way: initially, a layer of material is deposited on the substrate as described above, and it is shaped to its slice contour using a machining tool (like CNC). After machining, another layer of deposit is made, and this process of depositing layers by FS and shaping layers by CNC machining is repeated until the desired build height is achieved. This type of 3D part fabrication can be termed single-track FS. In addition to this, 3D part fabrication using FS can be done via a multitrack approach, too [16]. In this approach, initially, multitrack deposition is done using FS over the entire slice area by depositing layers that either partially overlap or are adjacent to each other, then shaping the complete area via CNC machining. Multitrack deposition of layers and shaping of layers via CNC machining steps are repeated until the complete part is built. The steps involved in AM of 3D components via FS are illustrated in Figure 5.8.

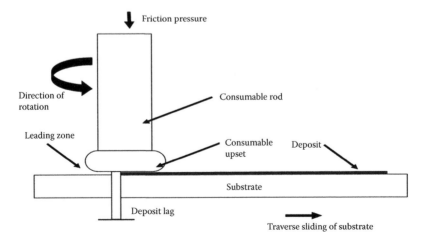

FIGURE 5.7
Schematic of friction surfacing.

Friction Deposition-Based Additive Manufacturing Techniques

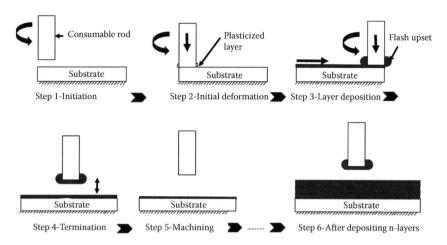

FIGURE 5.8
Steps involved in additive manufacturing of components via FS.

5.3.2 Friction Surfacing Process Parameters

The success of layer-by-layer deposition depends on the bond strength between layers and the thickness and width of deposits, which further depends on the process parameters involved during the FS process. Based on the published FS literature, it is quite challenging to generalize the individual effect of each process parameter. However, general trends based on different research outcomes are presented here in a simplified way. The main process parameters involved in FS are [17]:

a. Rotational speed of rod
b. Traverse speed
c. Axial pressure applied
d. Consumable rod diameter
e. Rod/mechtrode position

a. *Rotational speed*: The rotational speed defines the amount of heat generation. It is the general belief that with increasing rotational speed, enhanced friction increases heat generation, which in turn results in improved bonding efficiency. However, some published works have concluded that a combination of lower rotational speeds with low travel speeds results in an effective diffusion process and enhanced deposit widths. Thus, the concept of the effect of rotational speed needs attention and is discussed here. The effect of rotational speed on the heat input rate over the substrate and deposit/coating is quiet interesting. Shinoda et al. [18] reported the relationship between the heat input rate of the rod and substrate for different speeds of rotation as illustrated in Figure 5.9a. The rate of heat input of the deposit increases with an increase in rotational speed, while it decreases for the substrate. A similar trend can be seen in Figure 5.9b, which signifies that with increased speed of rotation, the HAZ area (heat input of area) of the substrate decreases. Similar results were reported by Rafi et al. [19]: with increased rotational speed, the width of the deposit decreases; however, the flatness of coatings increases and more regular deposits (H13 over mild steel in this particular study, described in Figure 5.10) are formed (refer to Figure 5.10). This can

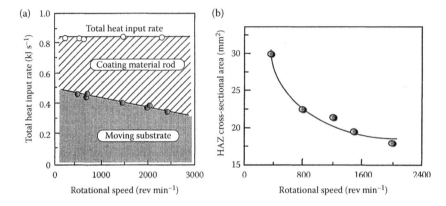

FIGURE 5.9
Effect of rotational speed on: (a) heat input rate of coating and substrate, (b) HAZ cross-sectional area of substrate. (Adapted from Shinoda, T. et al. *Surface Engineering*, 1998. 14(3): 211–216.) [18]

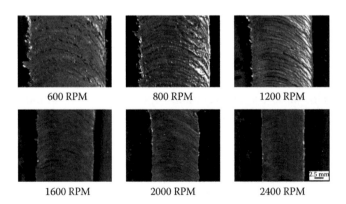

FIGURE 5.10
Macrographic image of influence of speed of rotation on roughness and width of coating of AISI H13 over mild steel substrate. (Adapted from Rafi, H. K. et al. *Surface and Coatings Technology*, 2010. 205(1): 232–242.) [19]

be attributed to the phenomena of decreased contact area between the substrate and coating material with the increased speed of rotation and consequent reduction in heat input transferred to the substrate material [20–22]. Thus, the rotational speed of the rod affects the bonding quality and width of deposits and can be utilized as a tool in customizing the heat inputs of the substrate and coating.

b. *Traverse speed*: The traverse speed of the substrate affects the rate of material deposition. A lower traverse (or feed) speed increases heat generation and results in higher rates of material deposition over the substrate. On the other hand, higher travel speeds produce thinner deposits by allowing lower heat exposure time. With lower heat exposure time, the process temperature decreases. Also, thinner deposits cool faster, and this combination result in less grain growth and fine microstructures, which further leads to enhancement of bond strength to some extent [23,24]. Thus, an increase in travel speed by holding other parameters constant would result in a reduction of the width and thickness of deposits. The effect of travel speed on the width and thickness of the coating material is shown in Figure 5.11 [25]. Thus, traverse speed affects the width, thickness, bonding area, and strength of coatings.

c. *Axial force*: Axial force is a critical parameter in FS. With an increase in axial force, the adhesion of the coating increases, while it decreases the thickness of the deposit and consequently increases the width of the deposit. However, with an increase in axial force, flash increases, which decreases the process efficiency [26]. Low axial force results in poor material

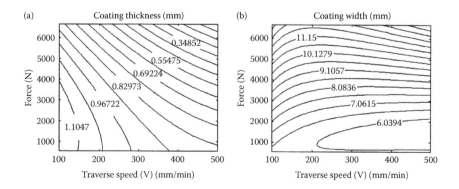

FIGURE 5.11
Effect of traverse speed on deposit variables. (Adapted from Vitanov, V. I. et al. *Surface and Coatings Technology*, 2001. 140(3): 256–262.) [25]

consolidation owing to insufficient friction heat and forging pressure. This can be seen in Figure 5.12, which illustrates the effect of variation of axial force at constant levels of rotational and traverse speeds during a deposit of mild steel over the same material. Thus, axial force should be adequately chosen to obtain sound deposits.

d. *Consumable rod diameter*: The diameter of consumable rod generally affects the width of deposit and heat exposure [16,17]. This implies that the larger the rod diameter, the wider the corresponding coating. However, a larger rod diameter demands higher equipment capability.

e. *Mechtrode position*: The significance of the position of the mechtrode is generally considered in multitrack friction surfaced (FSed) deposits. To date, no standard equation has been reported for the gap or overlapping of adjacent tracks. It should be manually adjusted based on prior experience, thickness of the track, diameter of the consumable rod, and type of material under investigation. However, improper positioning of the mechtrode leads to defect formation owing to improper heat generation at the interface of the previously deposited track and mechtrode tip. A typical good bonded and unbonded region obtained using proper and improper positioning of the mechtrode is as shown in Figure 5.13.

Thus, it can be concluded that the selection of optimum process parameters is a critical factor in defining the soundness of the fabricated component. Further, the effect of these parameters is strongly dependent on the combination of materials utilized for FS.

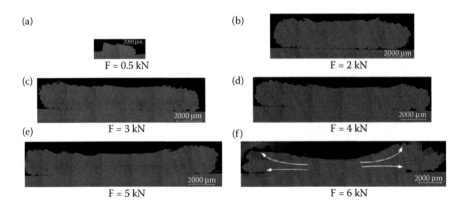

FIGURE 5.12
Effect of axial force on cross-section of coating of mild steel over same material. (Adapted from Gandra, J. et al. *Journal of Materials Processing Technology*, 2012. 212(8): 1676–1686.) [27]

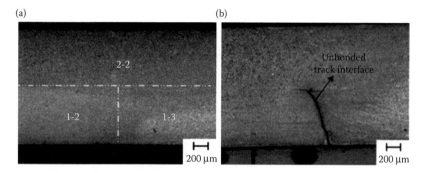

FIGURE 5.13
Optical micrographs of multilayer multitrack FSed deposits of mild steel; (a) good bonding at interface between layer 1-track 2, layer 1-track 3, and layer 2-track 2; (b) unbonded region between two tracks. (Adapted from Dilip, J. J. S. et al. *Materials and Manufacturing Processes*, 2013. 28(2): 189–194.) [16]

5.3.3 General Features and Status of Research of Friction Surfacing–Based Additive Manufacturing Methods

As discussed above, FS can be utilized as a solid-state AM tool for fabricating 3D parts with wrought microstructures possessing better mechanical properties as compared to BM. Such fabrication can be done via single- and multitracks. The fabrication of parts via a single-track scheme is simple, while using a multitrack scheme is quite interesting. In the latter case, the mechtrode position is manually controlled to obtain good bonding between different tracks, and the interface track bonding depends on the same. For the sake of more clarity over single- and multitrack depositions, some experimental results reported in the literature are discussed here. A study of multitrack, multilayer deposition of a mild steel consumable over a same-material substrate was reported by Dilip et al. [16]. Three layers were deposited in a brick-like fashion in order to unite different tracks in which the first layer consisted of five tracks, the second layer consisted of four tracks, and the third layer consisted of three tracks. The scheme of layers and track arrangement is as shown in Figure 5.14. The images of multitrack deposition and multilayer deposits are shown in Figure 5.14b and c, respectively.

FSed deposits reflect good bonding between different tracks and different layers, as depicted in Figure 5.15. The FSed deposit exhibits a fine-grained microstructure consisting of ferrite and pearlite, which is the typical microstructure of wrought mild steel (refer to Figure 5.15b). This is owing to the occurrence of dynamic recrystallization mechanisms during FS [16]. Due to excellent bonding between various interfaces and refined grain size, enhanced mechanical properties (YS: 260 MPa, UTS: 410 MPa, % elongation: 23) were achieved as compared to standard mild steel.

In addition to single-track and multitrack aspects, FS can be utilized for more interesting applications of producing 3D parts with fully enclosed

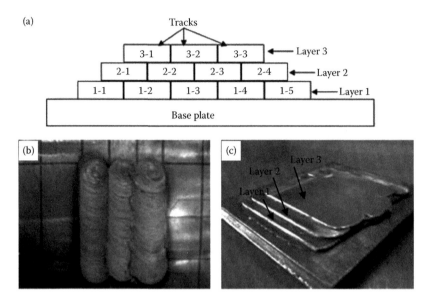

FIGURE 5.14
(a) Scheme of layer and track arrangement in FS deposit of three layers, in the numbering system, the first number defines layer and the second number denotes the track; (b) image of multitrack (three) deposits in first layer; (c) image of multilayer FS deposit. (Adapted from Dilip, J. J. S. et al. *Materials and Manufacturing Processes*, 2013. 28(2): 189–194.) [16]

cavities. Figure 5.16 illustrates the steps of the fabrication of a 3D object of AISI 410 martensitic stainless steel with three enclosed cavities, along with images of the intermediate and finished deposits [16]. An AISI 410 steel rod of 18 mm diameter was utilized as a consumable rod. Initially, five layers were deposited one over the other in single-track fashion, as shown in

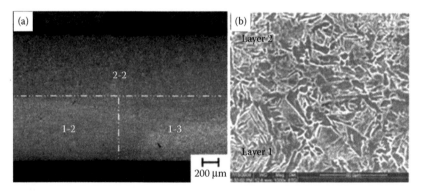

FIGURE 5.15
Microstructural images of multitrack multilayer deposit of mild steel showing bonding between different tracks and layers: (a) optical image of layer 1-track 2 and track 3 interface and layer 1 and layer 2 interface; (b) SEM image showing excellent bonding between layer 1 and layer 2. (Adapted from Dilip, J. J. S. et al. *Materials and Manufacturing Processes*, 2013. 28(2): 189–194.) [16]

Friction Deposition-Based Additive Manufacturing Techniques　　91

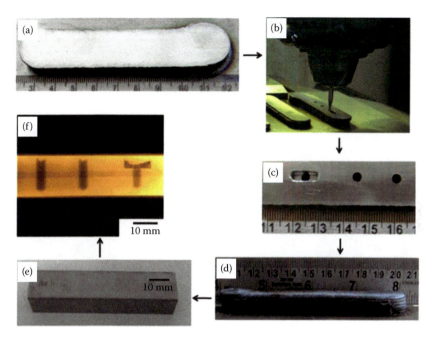

FIGURE 5.16
Steps and images of 3D part fabrication with fully enclosed cavities. (Adapted from Dilip, J. J. S. et al. *Materials and Manufacturing Processes*, 2013. 28(2): 189–194.) [16]

Figure 5.16a. Then, three holes were made in the deposit in the form of "I," "I," and "T" shapes (refer to Figure 5.16b and c). Subsequently, the sixth layer was deposited over the previously fabricated build with drilled cavities. The final deposit is as shown in Figure 5.16d. Figure 5.16e shows the finished object machined from the build with dimensions of 100 × 16 × 17 mm. The radiographic image of the build showing the longitudinal section with fully enclosed cavities is illustrated in Figure 5.16f.

Thus, from the successful results of the above work, it can be concluded that FS can be successfully utilized for fabricating components with fully enclosed features with different shapes and dimensions. The shape of the enclosed features is independent of the mechtrode, while the size of the enclosed cavities depends on the mechtrode diameter. A cavity that is too large for a defined size of mechtrode results in failure of the process, and the mechtrode tends to get struck in the cavity.

Another intriguing application of AM via FS is 3D component fabrication using different materials (two or more than two). Dilip et al. [16] reported experimental results on fabrication of different materials build using FS. Six layers were deposited in layer-by-layer fashion using AISI 316 and AISI 410 as consumables and were deposited in alternating layers over a mild steel substrate. The macrographic and microstructural images of the build are

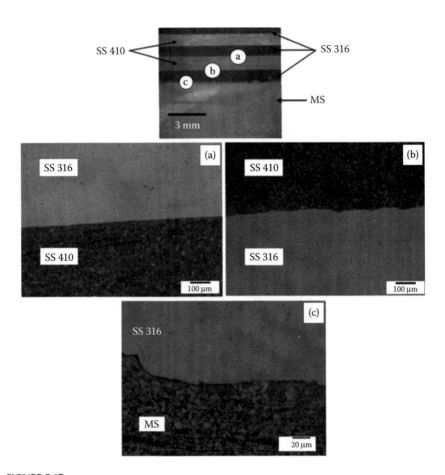

FIGURE 5.17
Optical micrographs of build fabricated by alternate deposition of alloy 316 and alloy 410 stainless steels via FS. (Adapted from Dilip, J. J. S. et al. *Materials and Manufacturing Processes*, 2013. 28(2): 189–194.) [16]

shown in Figure 5.17. Figure 5.17a and b show the interface between the 316 alloy and 410 alloy, while Figure 5.17c shows the interfacial region between the 316 alloy and mild steel. Upon microstructural observation, good bonding between different interfaces appears and no occurrence of deleterious phases was reported. The AISI 316 alloy exhibits high corrosion resistance and is a soft and ductile material, while the AISI 410 alloy is a hard and strong material with high wear resistance. The combination of these two alloys produces a hybrid material with favorably balanced properties of ductility, strength, wear, and corrosion resistance. The fabrication of such multimaterial objects using conventional AM techniques without harming the properties of these materials is quite difficult.

5.3.4 Benefits and Limitations

FS as an additive manufacturing tool has following benefits over conventional fusion-based MAM techniques:

- Being a solid-state process, it avoids various metallurgical solidification defects, which are common in fusion-based MAM.
- The achievable layer thickness and rate of deposition are higher in FS as compared to commercially available direct MAM.
- FS produces fine-grained microstructures. As FS is a hot plastic deformation process in which the maximum attainable temperature is much less than the temperature in fusion-based processes, part distortion can be prevented.
- FS can be utilized for a wide range of materials and also for dissimilar material combinations, which is generally difficult in conventional MAM processes.

The application of FS as an additive manufacturing process suffers some limitations, too. One of the major limitations is the machining of each layer in order to give it the desired shape, as poor bonding at the edges of the deposits takes place. Also, a revolving flash is generated at the rod tip, resulting in decreased mass transfer efficiency since it basically represents material unbonded to the substrate. FS provides restricted control of deposition thickness and breadth, which is due to the geometrical conditions of coatings being governed by a close process parameters range. Secondly, it needs to be ensured during FS that the build is not susceptible to free deformation/buckling due to process forces. The process works best for making parts with no downward-facing features or overhangs, though judicial process planning can troubleshoot this issue. Further, the characteristics of a given material that determine the ease with which it can be satisfactorily friction-surfaced on a given substrate are presently unclear [16].

5.3.5 Applications of Friction Surfacing as Additive Manufacturing Tool

In spite of FS being a solid-state depositing process, it is used as a suitable technique for AM of different materials. The common applications of FS include the following:

- An impressive application is manufacturing rib-on-plate structures similar to the ones used for aircraft frames. Forging and machining are the typical processes to achieve such structures, with a significant buy-to-fly cost (where large fractions of material purchased are removed as chips). Using FS, it is feasible to build up rib-on-plate structures possessing wrought-metal properties in an economic way [16].

- Depositing of a hard material layer to enhance wear-resistant coatings.
- Depositing hard material on agricultural equipment.

5.4 Conclusion

Based on experimental results over FD and FS, it can be concluded that both of these techniques can be successfully utilized as additive manufacturing tools for fabrication of 3D objects. These can be utilized on wide range of materials, including ferrous as well as nonferrous materials owing to their solid-state nature. The occurrence of severe plastic deformation and subsequent dynamic recrystallization makes these processes suitable for producing high-strength materials. FD can be used in producing similar materials and dissimilar materials, as well as composite deposits. FS can be utilized via single- and multitrack schemes. The application of FS in fabricating 3D objects with enclosed cavities is quite intriguing.

References

1. Dilip, J. J. S., Kalid R. H., Janaki Ram, G. D. A new additive manufacturing process based on friction deposition. *Transactions of the Indian Institute of Metals*, 2011. 64(1): 27.
2. Dilip, J. J. S., Janaki Ram, G. D. Microstructures and properties of friction freeform fabricated borated stainless steel. *Journal of Materials Engineering and Performance*, 2013. 22(10): 3034–3042.
3. Karthik, G. M., Ram, G. D. Janki, Kottada, R. S. Friction deposition of titanium particle reinforced aluminum matrix composites. *Materials Science and Engineering: A*, 2016. 653(Supplement C): 71–83.
4. Pérez, P., Garcés, G., Adeva, P. Mechanical properties of a Mg–10 (vol.%)Ti composite. *Composites Science and Technology*, 2004. 64(1): 145–151.
5. Thakur, S. K., Kong, T. S., Gupta, M. Microwave synthesis and characterization of metastable (Al/Ti) and hybrid (Al/Ti+SiC) composites. *Materials Science and Engineering: A*, 2007. 452–453(Supplement C): 61–69.
6. Hassan, S. F., Gupta, M. Development of a novel magnesium/nickel composite with improved mechanical properties. *Journal of Alloys and Compounds*, 2002. 335(1): L10–L15.
7. Yadav, D., Bauri, R. Processing, microstructure and mechanical properties of nickel particles embedded aluminium matrix composite. *Materials Science and Engineering: A*, 2011. 528(3): 1326–1333.

8. Karthik, G. M., Panikar, S., Ram, G. D. J., Kottada, R. S. Additive manufacturing of an aluminum matrix composite reinforced with nanocrystalline high-entropy alloy particles. *Materials Science and Engineering: A*, 2017. 679(Supplement C): 193–203.
9. Rathee, S., Maheshwari, S., Siddiquee, A. N. Issues and strategies in composite fabrication via friction stir processing: A review. *Materials and Manufacturing Processes*, 2018. 33(3): 239–261.
10. Dilip, J. J. S., Janaki Ram, G. D. Friction freeform fabrication of superalloy Inconel 718: Prospects and problems. *Metallurgical and Materials Transactions B*, 2014. 45(1): 182–192.
11. Dilip, J. J. S., Janaki Ram, G. D., Stucker, B. E. Additive manufacturing with friction welding and friction deposition processes. *International Journal of Rapid Manufacturing*, 2012. 3(1): 56–69.
12. Klopstock, H., Neelands, A. R. *An Improved Method of Joining or Welding Metals* 1945, GB 572789 A (ID: lens.org/168-899-190-400-639), Application: October 17, 1941 (GB 1339641 A), UK.
13. Bishop, E. Friction welding in the Soviet Union. *Welding and Metal Fabrication*, 1960. 408–410.
14. Nicholas, E. D. Friction Processing Technologies. *Welding in the World*, 2003. 47(11): 2–9.
15. Nicholas, E. D. Friction surfacing, in *ASM Handbook—Welding Brazing and Soldering* (Eds D. Olson, Siewert, T., Liu, S., Edwards, G.). 1993, ASM International: Ohio. pp. 321–323.
16. Dilip, J. J. S., Babu, S., Rajan, S. Varadha, Rafi, K. H., Ram, G. D. Janaki, Stucker, B. E. Use of friction surfacing for additive manufacturing. *Materials and Manufacturing Processes*, 2013. 28(2): 189–194.
17. Kramer de Macedo, M. L., Pinheiro, G. A., dos Santos, J. F., Strohaecker, T. R. Deposit by friction surfacing and its applications. *Welding International*, 2010. 24(6): 422–431.
18. Shinoda, T., Li, J. Q., Katoh, Y., Yashiro, T. Effect of process parameters during friction coating on properties of non-dilution coating layers. *Surface Engineering*, 1998. 14(3): 211–216.
19. Rafi, H. K., Janaki Ram, G. D., Phanikumar, G., Rao, K. P. Friction surfaced tool steel (H13) coatings on low carbon steel: A study on the effects of process parameters on coating characteristics and integrity. *Surface and Coatings Technology*, 2010. 205(1): 232–242.
20. Bedford, G. M., Richards, P. J. Paper 50, in *Proc. Surface Engineering Conference1985*, London, UK, TWI.
21. Fukakusa, K. On the characteristics of the rotational contact plane: A fundamental study of friction surfacing. *Welding International*, 1996. 10(7): 524–529.
22. Fukakusa, K. On real rotational contact plane in friction welding of different diameter materials and dissimilar materials: Fundamental study of friction welding. *Quarterly Journal of the Japan Welding Society*, 1996. 14(3): 483–488.
23. Rafi, H. K., Ram, G. D. J., Phanikumar, G., Rao, K. P. Friction surfacing of austenitic stainless steel on low carbon steel: Studies on the effects of traverse speed, in *World Congress on Engineering* 2010, London.
24. Kumar, V.B., Reddy, G. M., Mohandas, T. Influence of process parameters on physical dimensions of AA6063 aluminium alloy coating on mild steel in friction surfacing. *Defence Technology*, 2015. 11(3): 275–281.

25. Vitanov, V. I., Voutchkov, I. I. Bedford, G. M. Neurofuzzy approach to process parameter selection for friction surfacing applications. *Surface and Coatings Technology*, 2001. 140(3): 256–262.
26. Li, J. Q., Shinoda, T. Underwater friction surfacing. *Surface Engineering*, 2000. 16(1): 31–35.
27. Gandra, J., Miranda, R. M., Vilaça, P. Performance analysis of friction surfacing. *Journal of Materials Processing Technology*, 2012. 212(8): 1676–1686.

6
Friction Stir Welding-Based Additive Manufacturing Techniques

6.1 Introduction

Elaborations in Chapters 1 and 2 provide enough basis to assert that AM is one of the most promising technologies of modern times. It would not be out of place to envisage phenomenal growth in its application, provided concerns of fusion-based AM processes are suitably addressed. Most feasible solutions to limitations of fusion-based AM processes lie in solid-state AM technology. It is worthwhile to predict enormous research interest in solid-state AM processes. As discussed in the previous chapters, friction welding (rotary)–based and friction deposition–based AM methods are generally suitable for circular (rod form) objects. However, friction surfacing–based AM is suitable for the development of deposits in diverse shapes and profiles. In these techniques, the addition of material takes place either by friction joining of objects or deposition of material from a consumable rod. The concept of friction stir welding-based AM techniques is slightly different from these. Three processes are categorized in this class, which are, friction stir additive manufacturing (FSAM), friction assisted (lap) seam welding (FASW), and additive friction stir (AFS). In FSAM, the addition of material takes place in the form of joining (friction stir lap welding) of sheets/plates via a nonconsumable tool, whereas in AFS, the addition of material takes place via a hollow tool (filled with powder or a consumable rod). In FASW, the addition of layers/sheets takes place owing to deformation caused by a nonconsumable pinless tool and applied axial pressure. These techniques can be commonly termed friction stir additive techniques (FSATs). FSATs are also a special class of friction based additive techniques. These FSATs can suitably address issues encountered during conventional MAM [1–3]. FSAM is capable of customizing microstructures of parts and can considerably improve the structural efficiency of fabricated components. Similarly, FASW produces parts with superior mechanical properties as compared to their parent metals. Aeroprobe's AFS/AFS is a MAM technique with improved speed and volume that enables low-cost, fully dense components which closely comply with the dimensional requirements. AFS produces wrought components from a broad range of materials. The majority of the research is focused on intricate shapes, accuracy, and materials. It is required that more emphasis of ongoing research should be directed in developing components that can meet strict strength restrictions of user industrial sectors, such as transportation. These issues cannot be suitably addressed without developing a thorough understanding of FSW, which is the parent process of FSAM, FASW, and AFS. A brief discussion of FSW is presented in the subsequent section.

6.2 Friction Stir Welding

Friction stir welding is a customized version of the conventional solid-state FW process. It is broadly based upon the fact that frictional heat will be generated if two surfaces are rigorously rubbed with respect to each other, which raises temperatures below the melting point temperature of materials. At this temperature, the material gets softened and the rotational and traverse motion of the tool leads to mixing and consolidation of material that forms a joint at the interface. FSW was patented by The Welding Institute in the United Kingdom in 1991 [4]. A significant milestone in the invention of FSW was success in joining unweldable aluminum and its alloys. Before the advent of FSW, some series of aluminum alloys, like the Al 2xxx and 7xxx series, were either unweldable or extremely difficult to weld with the help of conventional welding techniques [5]. Although a few conventional fusion-welding techniques were in use for their welding, corresponding weld strengths were still quite poor owing to overaging and increased grain size at elevated temperatures.

The working principle of FSW is very simple. In its conventional form, a nonconsumable rotating tool with a specially designed pin and shoulder plummets into the joint line of base metal plates (in case of butt joining) while traversing in a predefined direction to cover up the desired realm. The process schematic is shown in Figure 6.1. Softening and plasticization of the work material occur owing to local frictional heat generation between the tool and work pieces. The stirring action of the tool mixes the heated material, which is subsequently consolidated by forging action behind (under) the tool shoulder. As the tool/BM traverses, joint formation succeeds the tool motion. Different types of joint formation can be done in FSW depending on their application and the flexibility of FSW machines. The most commonly used joint configurations for FSW are butt and lap joints. In a butt joint, generally two plates with identical dimensions in terms of thickness and length are tightly fixed in a fixture to prevent the parting of the plates during tool pin insertion. In this configuration, abutting edges form the joint, as shown in Figure 6.2a. In the case of a lap joint, overlapping sheets/plates are joined (refer to Figure 6.2d–f). From Figure 6.2, it is clear that with the application of

Friction Stir Welding-Based Additive Manufacturing Techniques

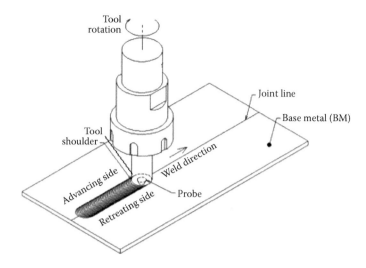

FIGURE 6.1
Schematic diagram of FSW. (From Siddiquee, A. N., Pandey, S. *The International Journal of Advanced Manufacturing Technology*, 2014. 73(1): 479–486.) [6]

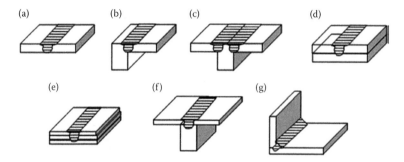

FIGURE 6.2
Joint designs for FSW: (a) simple butt joint, (b) edge butt joint, (c) T-butt, (d) lap joint, (e) multiple lap, (f) T-lap, (g) fillet joint. (From Mishra, R. S., Ma, Z. Y. *Materials Science and Engineering: R: Reports*, 2005. 50(1–2): 1–78.) [7]

lap and butt joints, more joint configurations can be made to fulfill specific engineering applications using FSW.

Pressure and plastic deformation play a major role in FSW. Furthermore, the heat-affected zone in FSW is narrow due to low and confined heat input as compared to fusion-welding processes, which leads to finer grains and higher mechanical properties of the joints. Owing to these benefits, FSW has emerged as one of the best options in solid-state joining techniques especially in welding of aluminium and other light weight metal alloys. The first industrial application of FSW was around 1995. This technology has progressed leaps and bounds since then and has proven strength as an innovative enabling joining strategy. It offers a wide spectrum of applications ranging from defense, naval, aeronautical, automobiles, and so on. Besides its core welding applicability, lots of FSW variants like friction stir butt welding (FSBW), FSLW, spot friction stir spot welding (FSSW), friction stir processing (FSP), and so on are also finding utility. Another very innovative and interesting application of FSW is additive joining of sheets/plates. Three FSW-based AM methods are in practice, which are, FSAM, FASW, and AFS, which are utilized for fabricating metallic layers using layer-by-layer AM principles. The commonly used friction stir–based conventional and hybrid processes are summarized in Figure 6.3.

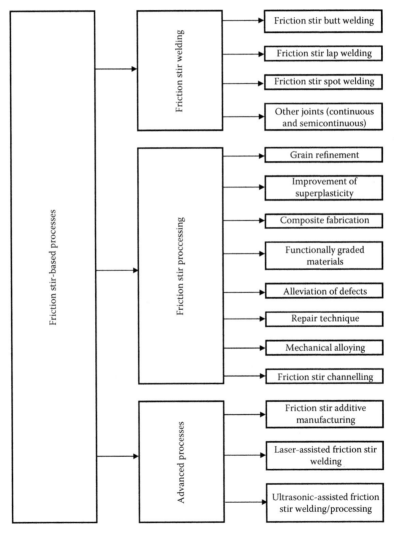

FIGURE 6.3
Friction stir processes based on their working and applications.

6.2.1 Terminology Used in Friction Stir Welding

Various FSW terminological concepts illustrated in Figure 6.1 and other related aspects are defined in detail in this section.

Tool rotational speed: The speed at which the tool rotates either in the clockwise or in anticlockwise direction is termed the tool rotational speed. It is generally measured in rotations per minute (rpm) or rotations per second (rps). Tool rotational speed has a major impact on the material flow and heat input during the process [8,9].

Tool traverse speed: This is also called tool travel speed or welding/processing speed. It is the speed at which the rotating tool moves in the forward direction longitudinally. It is generally measured in cm/minute, mm/minute, or mm/second. It affects the heat input per unit processing length and also contributes to the process thermal cycle.

Tool tilt angle: This refers to the inclination of the spindle axis normal to the base material surface or the angle of the spindle from its vertical position. It is measured in degrees. Generally, an angle of 0 to 3 degrees is selected during FSW/P [10].

Tool shoulder: The portion of the tool between its periphery and the root of the pin that comes in direct physical contact with the workpiece during the process is known as the tool shoulder.

Tool pin: This is also termed the tool probe. It is the lowest part of the tool adjacent to the shoulder. There can be different types of tool pin profiles, like cylindrical, conical, square, triangular, threaded cylindrical, and so on. In addition to this, different shapes of pins have evolved.

Advance/advancing side: The side at which the direction of rotational speed is same as that of the tool travel speed is called the advancing side (AS).

Retreating side: The side at which the tool rotational speed direction is opposite that of the tool travel speed direction is called the retreating side (RS). At the RS, the material flow is smoother as compared to the advancing side, as the rotation of the tool supports the backward flow of the material.

Leading edge: The frontal tool surface that comes in contact with the cold workpiece material is termed the leading edge.

Trailing edge: The rear side of the tool surface is termed the trailing edge.

Axial force: The material flow due to extrusion and forging gives rise to a vertical tool force at the interface of the shoulder and the material being processed. Its magnitude depends on the properties of the material, such as its strength and the softening temperature that prevails during welding. This force is responsible for holding the workpiece and material during flow (during FSW/P).

Tool plunge depth: The shoulder insertion depth below the base metal's free surface is termed as tool plunge depth. It is a critical parameter in FSW/P.

Z-axis: The axis perpendicular to the job is termed as the z-axis. Axial force also acts along it.

X-axis: The tool travel direction is considered the x-axis.

Y-axis: The y-axis is normal to the traverse direction and generally parallel to the top job surface.

The above terminology described for FSW is similar for FSAM, FASW, and AFS.

6.3 Friction Stir Additive Manufacturing

FSAM is a solid-state additive manufacturing process that can be considered a confluence of layer-by-layer AM and FSW processes. FSAM marked its first commercial presence with the technical reports published by Airbus as early as 2006 [11]. However, not much attention was paid at that time to this process. After that, Boeing introduced FSAM as a technique for developing energy-efficient structures [12]. Research in the field of FSAM has, however, been active since 2014. Several authors [1,13,14] acceded to the fact that it possesses a unique ability to modify and considerably improve the microstructures of the fabricated components. Owing to its unique ability to engineer and control microstructures, it can help in tailoring microstructures to suit the customer's requirements. It results in fine-grained structures, which in turn renders significantly improved ductility and structural performance to the fabricated components in comparison to the base materials. It can be successfully employed in fabricating high-strength metallic alloys [1,13,15].

6.3.1 Working Principles of Friction Stir Additive Manufacturing

FSAM utilizes the principle of layer-by-layer AM. Friction stir lap welding is generally utilized to additively join the metallic layers. In its conventional form, a nonconsumable tool is inserted into a stack of overlapping sheets/plates and FSLW is carried along the defined direction with optimum process parameters. A schematic diagram for the FSAM process is shown in Figure 6.4.

The heat necessary for the joining of layers is obtained due to friction and plastic deformation of the work material [1]. Joint creation takes place due to heat generation, material movement and consolidation of material from the front to back portions of the tool. The resulting layer thickness is dependent on the following: (a) thickness of each sheet/plate additively manufactured, (b) assembling methodology, (c) number of individual layers/sheets, (d) process parameter restrictions like the dimensions of the tool pin. The geometrical design of the tool during FSAM controls the macroscopic and microscopic aspects of the fabricated build/stacks in addition to process parameters.

6.3.1.1 Steps Involved in Friction Stir Additive Manufacturing

The basic steps involved in FSAM are presented in Figure 6.5 and can be described as:

Friction Stir Welding-Based Additive Manufacturing Techniques

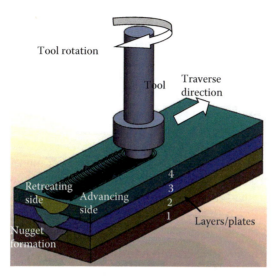

FIGURE 6.4
Schematic arrangement of FSAM.

Step 1. Initial preparation:

First of all, the plates/sheets to be additively manufactured using FSAM are prepared in terms of flatness. These plates are made in the proper dimensions as desired and cleaned with acetone.

Step 2. Stack metal sheets:

Initially, two plates/sheets are overlapped one over the other as per the orientation of the desired build.

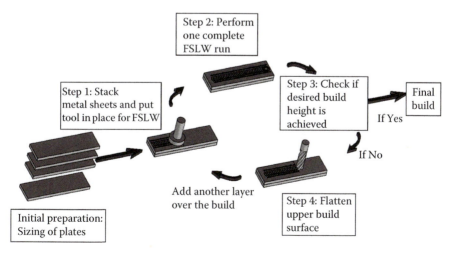

FIGURE 6.5
Steps utilized in FSAM.

Step 3. Perform one complete FSLW run:

After the placing of the two plates/sheets, FSLW is performed with suitable process parameters. After completion of the first run, if the desired build height is achieved, then it is the final component. Otherwise, the process progresses to step 4.

Step 4. Flatten upper build surface:

If the desired build height is not achieved, additional layers are deposited over the build. For this purpose, the upper processed surface of the previously fabricated build is flattened by some means (e.g., by machining) to remove the flash developed during FSLW. After this, a new plate/sheet is placed over the flattened build and steps 2–4 are followed until the desired build height is achieved.

6.3.2 Friction Stir Additive Manufacturing Process Variables

The process variables of FSAM are similar to those of FSW/P. The main process variables and their effect on component properties and process efficiency are discussed here.

The rotational and travel speeds of the tool are the major process parameters affecting heat generation and the properties of the consolidated material. Higher tool rotational speeds result in improved heat generation. It is well established that a combination of higher rotational speeds and lower travel speeds increases heat input and improves material flow. However, this combination also depends on other parameters like tool plunge depth, tilt angle, tool shoulder diameter, tool geometry, and so on. Also, different types of materials demand different optimum heat values. For example, aluminum alloys demand less heat for joining as compared to steels. Tool plunge depth and tool tilt angle are other main process variables that affect material joining by aiding heat generation [16]. This can be simplified as: at a fixed tool plunge depth, a decrease of the tool tilt angle increases the contact area of the tool and workpiece, resulting in greater heat generation. Thus, optimum selection of tool plunge depth and tilt angle is needed for adequate joining. Axial force is another important parameter for FSAM and FSW/P processes. A lower axial load leads to poor consolidation of plasticized materials, resulting in tunnel defects. The main spindle head and bearings supporting the tool should be robust enough to support sufficient loads.

Other major variable are categorized into tool variables. Tool-specific variables also affect the heat generation and material flow during FSAM [1]. Tool variables mainly include tool geometry and tool dimensions. Further tool geometry involves the design of the tool shoulder undersurface and tool pin profile. A simple design of the shoulder end surface consists of a flat surface normal to the axis, or it can be concave or convex (i.e., at an angle to the tool axis). In simple terms, tool shoulder design consists of the shoulder diameter and angle of the end surface of the shoulder. A flat surface normal to the tool axis is the simplest design, but in some cases, it may cause excessive flash during FSW/P, which can be avoided using a concave shoulder. Additionally, the end surface may also have features such as scroll, spiral, or concentric circular grooves to enhance the performance of tools [17]. Tool pin design mainly consists of pin diameter, pin length, and the shape of the pin. The geometry of the tool pin governs the material flow. Numerous types of tool pin profile (such as cylindrical, conical, prism/pyramids of triangular or square cross-sections, lobular, or triflute) are in common use for FSW/P. A detailed discussion of the effects of these pin profiles is out of the scope of the current chapter.

6.3.3 General Features and Status of Research

6.3.3.1 Grain Size Variation

FSAM can basically be understood as involving repeated/multi-FSLW. During the addition of multisheets/layers, interesting microstructural characteristics of different layers are obtained owing to the complexities involved in various regions from the bottom to the top layer. It is now well accepted that FSAM is a customized version of FSW, which in itself is a severe plastic deformation process. Dynamic recrystallization during FSW produces equiaxed and refined granular microstructure in the stir zone of the processed region for various alloys, including aluminum alloys [18–20], copper alloys [21,22], magnesium alloys [23–25], steel [26,27], and so on. Thus, during FSAM, material is subjected to vigorous stirring and plastic deformation followed by DRX [13]. On the other hand, the material being processed experiences annealing due to the high process temperature. Here, in FSAM, the effect of annealing is dominated by the DRX effect [1]. Although FSAM produces fine-grained structures, there also exist large variations in grain sizes in different regions of the fabricated build. This variation in grain sizes can be attributed to different thermal exposures. For example, suppose "n" layers are to be additively manufactured by FSAM. According to the procedure, initially, the first two layers are additively joined, during which these layers experience a certain degree of thermal exposure. In the next stage, when the third layer is added over the previously fabricated build, it experiences thermal exposure for the first time, whereas the previously built layers are already preheated and experience thermal exposure for the second time. Similarly, when the fourth layer is added, it experiences thermal exposure for first time, while the third layer experiences it for the second time and the first two layers for the third. In this way, during the joining of n layers, the first two layers experience n − 1 thermal exposures, the third layer n − 2, and so on. During the entire processing, each layer experiences a different range of temperatures, and after a certain thickness, the effect of thermal exposure for the initially built layers decreases and there may be no negative effect of the process temperature. Thus, it can be concluded that every time a new layer is added, all the previously built layers are affected by different levels of temperature. Owing to these multiple and different thermal exposures for each layer or certain regions, microstructural features also reflect a difference in different layers. Another important reason for variation in grain sizes and microstructural characteristics may basically be attributed to the different types of material flow and strain induction at various build layers. For example, material flow near the top layer is shoulder driven, whereas at the bottommost layer, is it is pin driven. Thus, the combined effect of different heat exposures, material flow, and strain induction results in large variation in grain sizes at different locations of the build. Further, there is a dependence on the type of material being processed. This is explained here with the help of two examples taken from experimental studies published in the FSAM literature. One set of experimental results was reported by Palanivel et al. [1] in fabrication of magnesium-based WE43 alloys. They reported that the material flow near the top layer (which in this case is layer 4) is shoulder driven and higher strain is induced, resulting in a large reduction of grain sizes (refer to Table 6.1). On the other hand, material flow at the bottom layer was pin driven (layer 1) and therefore subjected to comparably less strain, resulting in larger-than-average granular sizes. Thus, from this study, it can be concluded that grain size in the top layer is smaller than that at the bottom layer, and it increases from the top to bottom layer in thickness. In another study, Yuqing et al. [13] reported fabrication of nine layers of an AA7075 build using FSAM. They observed an increased grain size of layers from top to bottom in thickness. That means the topmost layer exhibited smaller grains as compared to the bottom layer. This can chiefly be attributed to the fact that when additional layers are added to the previously formed build, the previously added layers are subjected to repeated thermal exposure and static annealing, due to which recrystallized grains and precipitates tend to elongate in heat-treatable AA7075 alloys. However, a larger grain size at the top layer surfaces as compared to the center of the same layer may be

TABLE 6.1

Grain Sizes at Different Locations of WE43 Magnesium Alloy Build Fabricated Using FSAM

Location	Grain Size, Average (μm)	Grain Size Maximum (μm)	Grain Size Minimum (μm)	Standard Deviation	Variation Coefficient
Layer 4	0.7	5.5	0.28	0.68	0.97
Interface 3	0.98	6.2	0.26	0.83	0.85
Layer 1	0.84	3.8	0.29	0.54	0.64
TMAZ	1.1	14.9	0.3	1.04	0.95

Source: Palanivel, S. et al. *Materials & Design (1980–2015)*, 2015. 65: 934–952. [1]

observed due to coarsening of grains at elevated temperatures [13]. From the results of experimental studies, it can be concluded that:

- Variation in grain sizes takes place in different regions of the build.
- This variation is dependent on the type of material (heat treatable or non–heat treatable) under investigation.
- The main reasons behind these variations are multiple and different thermal exposures, variation in material flow, and strain induction.
- The top layer is shoulder driven, while the bottom layer material flow is pin driven.
- Intermediate layers are subjected to shoulder-driven material flow when they are at the top of the stacking region, while they experience pin-driven material flow when another layer is added onto it. Thus, intermediate layers experience both material flow types.

6.3.3.1.1 Strengthening mechanism of components fabricated via FSAM

Strength represents resistance to deformation of the material, manifested in the form of dislocation movement, which further depends on dislocation density, total grain boundary area, presence of foreign/solute elements, and secondary phases in material. The primary strengthening mechanism responsible for enhancing strength in FSAM components is the Hall-Petch effect [13]. The kinetics of microstructural features in metallic components fabricated via FSAM depends upon the relative location and the impact of various strains along with the thermal cycle. Preferable kinetics obtained during FSAM amounts to highly desirable mechanical and fatigue characteristics. Extremely high formability and superplastic strength are achieved as a result of equiaxing, recrystallizing, and fine-grained microstructures [1]. Owing to thermal exposure, severe plastic deformation and recrystallization development of fine equiaxed grains in the stir zone of FSAM components takes place. This grain refinement effect leads to increased microhardness of components according to the Hall-Petch equation [9,13,28,29]. As a consequence, there is also an increase in the tensile strength of the fabricated build. In a particular study, high-strength WE43 magnesium alloy components were fabricated via FSAM that possessed far superior strength and ductility as compared to the base alloy [1]. An enhancement of ~23% in microhardness was achieved in fabricated build as compared to BM. The variation of mechanical properties of the fabricated build and BM can be seen in Figure 6.6. Figure 6.6a and b show a comparison between the microhardness of two builds fabricated at different rotational speeds in as-fabricated and aged conditions with the base alloy. The vertical dashed line in both figures depicts the microhardness of the rolled base alloy. It is evident from the figure that the FSAM fabricated build displays considerable improvement in microhardness as compared to BM. Figure 6.6c shows the enhancement in tensile strength of the fabricated build as compared to the base alloy. In another study, Palanivel et al. [15] successfully fabricated an

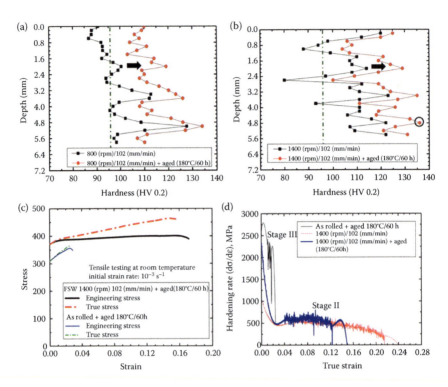

FIGURE 6.6
Microhardness readings along the direction of build at two different rotational speeds; (a) 800 rpm/102 mm/min, (b) 1400 rpm/102 mm/min, (c) tensile properties of build at 1400 rpm/102 mm/min and base metal after aging, (d) work-hardening behavior of as-received and fabricated build. (From Palanivel, S. et al. *Materials & Design (1980–2015)*, 2015. 65: 934–952.) [1]

AA 5083 build via FSAM in which the microhardness was enhanced by 18.2%. Further, considerable improvement in the yield strength of the fabricated build was also observed. The mechanical properties of the fabricated build and BM for an aluminum 5083 alloy, along with the macrographic image, are shown in Figure 6.7. Similar types of results were reported by Yuqing et al. [13] in which FSAM was utilized for fabrication of a multistack AA 7075 build. The tensile strength of all the fabricated layers was increased. Thus, microstructural reformation can be accounted for by the strengthening mechanism during the FSAM process.

6.3.4 Defects Associated with Friction Stir Additive Manufacturing

In its conventional form, FSAM is a kind of multiple-friction stir lap welding (FSLW) process. Similar to FSLW, different kinds of defects are associated with FSAM owing to improper selection of process parameters. The defects associated with FSAM are similar to FSLW defects. The most common defects include hook formation, cavities, cracks, kiss bond formation, and so on.

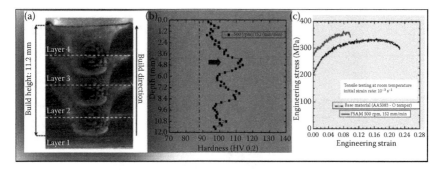

FIGURE 6.7
(a) Macrographic images of FSAM-fabricated build of Al 5083 alloy; (b) hardness comparison of base metal as-built components; vertical dashed line shows microhardness of base metal; (c) stress–strain curves for FSAM-fabricated built and base metal. (From Palanivel, S. et al. *JOM*, 2015. 67(3): 616–621.) [15]

A hook is a major defect that generally occurs in FSAM components and FSLW. The mechanism of hooking formation can be understood as: during dwell time, the tool pin first penetrates into the upper plate and subsequently into the lower plate while the tool shoulder remains inserted into the upper plate surface only. The rotation of the tool amounts to the propulsion of material from the lower plate toward the upper plate by the tool pin. Unbounded regions are formed at the lap weld extremities during this period. These unbounded regions also consist of partially bonded areas with trapped material that is present in overlapping sheets in contact. These partially bonded areas have generally curved regions and distorted interfaces and are known as hook defects [30,31]. The direction of the hook bend is generally the same as that of the material flow. In addition, the hooking defect may occur on both sides (i.e., AS and RS) of the stir zone; however, its shape on the AS is sharper as compared to the RS [32]. Its deflection generally varies from 0 to 90° [33]. The occurrence of the hook defect generally takes place in the thermomechanical affected zone, that is, the region surrounding the weld nugget. The chances of hooking at the heat affected zone are negligible, as these zones only experience heat exposure of welding [34]. A typical hook defect in an AA 7075 build fabricated via FSAM is shown in Figure 6.8.

Void is another major defect that generally occurs during FSAM. It is basically a voluminous empty space possessing no material aligned to the FSW/P direction [35]. This is also a special kind of surface defect and generally takes place at the advancing side due to inadequate heat and pressure. Due to reduction of pressure at the AS, the pin velocity reaches its minimum value, which in turn results in voids and tunnel defect formation [36–38]. A tunnel can be regarded as a continuous hole throughout the processing length beneath the top surface. Reduced heat generation occurs because of lower speeds of rotation and higher traversing velocities [39]. Further reduction in the amounts of generated heat leads to the formation of wormholes [40,41]. Another common defect is the occurrence of pores. Pores are relatively small-diameter

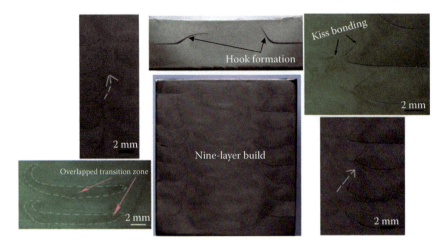

FIGURE 6.8
Defects in a Al 7075 alloy component fabricated via FSAM. (From Yuqing, M. L. et al. *The International Journal of Advanced Manufacturing Technology*, 2016. 83(9): 1637–1647.) [13]

holes of about 0.1–0.5 mm diameter. Lower tilt angles and reduced penetration depths are the most prominent reasons for pore formation [42,43]. Inadequacy of parameter selection, like too-high traverse speed, low rotational speed, improper tool pin offset, tilt angle, and forging pressure, is the major reason for the occurrence of voids, tunnels, and pores. Kissing bonds are another main class of defects that occur in component manufactured using FSAM [13]. This bonding defect occurs at interfacial regions in the build stir zone. Inadequately stirred and insufficiently deformed faying surfaces are the chief cause of improper bonding owing to the presence of an oxide layer over the interfaces [44]. In addition to this, inadequate heat and improperly mixed/deposited material can also result in kissing bond formation. The mechanical properties of the fabricated build are seriously affected by inadequate/poor bonding between faying surfaces due to the presence of kiss bonds [13]. A typical kiss bond defect is shown in Figure 6.8.

6.4 Friction Assisted Seam Welding-Based Additive Manufacturing Method

The utilization of the friction-assisted (lap) seam welding method for AM of metallic parts was proposed and patented by Kalvala et al. [3] in 2014. This method can also be performed on a conventional friction stir welding machine. The basic operation of the FASW method appears to be quite similar to FSAM at the first instance. However, there is a difference in working between these methods. In FSAM, a nonconsumable rotating tool with a designed pin and shoulder is inserted into the upper surface of overlapping sheets/plates. The tool pin then axially advances into the lower layer/plate up to a predetermined level (according to the tool pin height), whereas the shoulder slightly inserts (according to a defined plunge depth) into the upper surface of the top plate. On the other hand, in the FASW method, a nonconsumable rotating tool without a pin touches the upper surface of the top layer, an axial pressure is applied, and dwell time is allowed in this condition [3,45]. Another difference lies in the fact that, in FSAM, material undergoes plastic deformation and intermixing owing to the stirring action of the tool pin and shoulder, while in the latter case, localized deformation of material takes place owing to the rubbing action of the shoulder tool only, stick-slip phenomena, and applied axial pressure.

6.4.1 Working Principles of Friction-Assisted Seam Welding

Initially, two plates/sheets/layers of materials are placed overlapping each other [3,45]. The pinless rotating tool (one end fixed in the spindle of the machine) is allowed to advance in the axial direction to touch the upper surface of the top layer. The rubbing action of this tool generates frictional heat. Due to frictional heat generated at the interface of the upper surface of the top layer and the tool end, this surface undergoes plastic deformation, with further application of axial pressure up to the stage where the upper as well the bottom surface of the top layer and the upper surface of the bottom layer get deformed. The localized temperature increases at the interfaces, which may be the result of one or more of the mechanisms, that is, conduction of heat (between the tool end and the material surface), continuous stick-slip phenomena, and plastic deformation. At this stage, metallurgical bond formation takes place, which may be resultant of one or more mechanisms, that is, formation of fine-grained structure (due to recrystallization mechanism), stick-slip phenomena (which breaks oxide films between mating surfaces), and diffusion between interfaces of metallic sheets (refer to Figure 6.9). By allowing traverse motion up to a desired length, a continuous seam weld can be formed. The schematic of this method is shown in Figure 6.10. The above-mentioned principle is defined for single-track seam welding. Following similar steps, multitrack seams welding can be performed. For sound welds in multitrack joining, the second track overlaps to first up to a level of about 10%–20% of the width of the first processed region, and this process is continued until coverage of the entire desired width [3]. For developing 3D components and realizing AM principles, multiple layers are seam-welded via the principle described above using a single-track or multitrack scheme.

6.4.2 Status of Research and Recent Developments

The FASW method for AM of metallic components has recently been proposed [45], and not much information is available in the literature about this novel

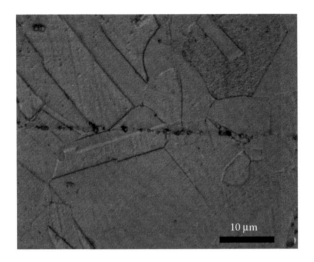

FIGURE 6.9
Diffusion phenomena showing extension of grains from one metallic layer to other in a typical AISI 304 friction seam weld. (From Kalvala, P. R. et al. *Defence Technology*, 2016. 12(1): 16–24.) [45]

proposed method. However, Kalvala et al. [45] reported seam welding of several materials like AISI 304, Ti6Al4V, Al 6061, and so on for similar as well dissimilar material combinations in a thickness range of 0.5–5 mm. Typical single-track seam welds of AA6061 and a multitrack multilayer seam weld of AISI 304 are shown in Figure 6.11a and b, respectively. The optimum parameters of 800 rpm tool rotational speed, 20 mm/min travel speed, and 5 kN axial force were utilized to perform the AA6061 seam weld, while 1600 rpm rotational speed, 15 mm/min travel speed, and 8 kN axial force were utilized to perform the multitrack seam welds of AISI 304 alloy.

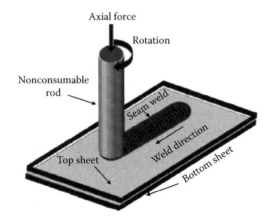

FIGURE 6.10
Schematic of FASW method. (From Kalvala, P. R. et al. *Defence Technology*, 2016. 12(1): 16–24.) [45]

FIGURE 6.11
Friction assisted (lap) seam welds: (a) single seam weld of AA6061 (top view); (b) multilayer multitrack weld of AISI 304: (1) top view, (2) cross-sectional view, (c) cross-section. (From Kalvala, P. R. et al. *Defence Technology*, 2016. 12(1): 16–24.) [45]

The results of shear tensile tests of different materials in this study reflect that most of the fabricated seam welds show enhanced tensile strength (except AA6061) as compared to their respective BMs. These results are listed in Table 6.2. The enhancement of strength can be owing to good bonding between overlapped sheets and refined grain size at the weld zone. The lower strength in AA6061 seam welds is basically due to the annealing effect experienced by heat-treatable AA6061.

Thus, from results of experimental studies [3,45], it can be concluded that the FASW method can be successfully utilized as an AM tool for producing 3D objects from a wide range of metallic alloys. Similar as well as dissimilar multilayer metallic builds can be developed via single- as well as multitrack schemes.

TABLE 6.2
Results of Tensile Test During FASW

Material under Investigation	Tensile Strength (MPa)	
	BM	Seam Weld
AA6061	256	158
C-Mn steel	375	390
AISI 304	521	580
HX	762	783

Source: Kalvala, P. R. et al. *Defence Technology*, 2016. 12(1): 16–24. [45]

6.5 Additive Friction Stir Process

AFS is purely additive in nature as compared to other FATs that involve additive and subtractive operations. A wide spectrum of raw materials can be used for AFS, as in this case, powders, metallic sheets/plates, tubes, and so on can be used as initial materials. Fabricated components possess wrought homogeneous microstructural properties [46,47]. This technique basically refines grain sizes and thereby imparts improved ductility and toughness characteristics to the components. Refinement in grain size occurs because the material is subjected to severe plastic deformation and new equiaxed smaller grains are formed by the recrystallization mechanism. However, these accentuate the already critical anisotropic behavior and are therefore undesirable [48]. AFS produces near net-shaped components by directly depositing material and has established itself as one of the most effective means of grain size refinement while simultaneously homogenizing microstructures and eliminating pores [49].

6.5.1 Working Principles of Additive Friction Stir

During AFS, material addition in the form of metal powder or a solid rod takes place from the center of a nonconsumable tool and flows in random directions owing to tool traverse motion. Tracks for each layer are overlapped, and each subsequent layer is deposited over the predeposited layer. After deposition of a layer over the substrate, the tool height is adjusted to accommodate the deposition of the subsequent layer. Fillers may be induced in previous layers to achieve enhanced bonding. Due to frictional heating and plastic deformation owing to rotation of the tool with regard to the substrate, a strong metallurgical bond forms between successive layers [50]. A schematic of this process is illustrated in Figure 6.12.

6.5.2 Microstructural Characterization in Components Developed via Additive Friction Stir

As previously discussed, AFS produces fine equiaxed grain structures with good metallurgical bonding [50]. The deposited material/build is normally

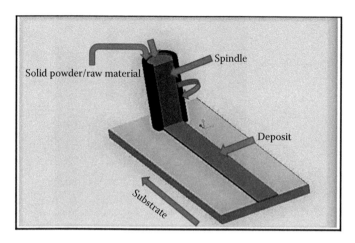

FIGURE 6.12
Schematic arrangement of AFS.

free from common defects of fusion-based MAM techniques like porosity, dilution, hot cracking, and so on. The chief contributing reason behind this is the solid-state material deposition. Process temperature in AFS generally lie between 0.6 T_m and 0.9 T_m, where T_m is the melting point temperature of the material. These are similar to the working temperatures in the case of FSW/P [7]. At such temperatures, the stirring action of the tool plastically deforms the deposited material. After this stage, recrystallization occurs resulting in fine-grained microstructures and induction of various desirable properties in the resultant fabricated components. This is illustrated here via an example. Figure 6.13a and b show the optical microstructural images of as-received IN625 material and AFS build IN625, respectively. It is evident from these images that there is a remarkable refinement of grains after the AFS process. The as-received material has an average grain size of ∼30 μm, while the AFS-fabricated Inconel 625 has a considerably reduced grain size of 0.26–1.3 μm (refer to Figure 6.13c–e) [50]. In addition to this, the results of high strain rate testing reflect an enhancement of ∼200 MPa in engineering strength over the corresponding quasistatic results.

In another study, WE43 Mg-based alloys were fabricated using both powder and plate as fillers to create multilayered structures by Calvert et al. [47]. Three samples were fabricated in which two samples were developed using atomized WE43 Mg-based alloy powders and a third sample was fabricated using a filler rod of WE43-T5 material. The first two samples were fabricated at same process parameters of 450 rpm rotational speed and 127 mm/min traverse speed with different cooling mediums (in first sample, water cooling was used, while in the second sample, nitrogen cooling was utilized). The third sample was fabricated at 800 rpm and 102 mm/min, and liquid nitrogen was used as cooling medium. It was reported that considerable refinement of grains in the deposited region occurred. However, the grain refinement in

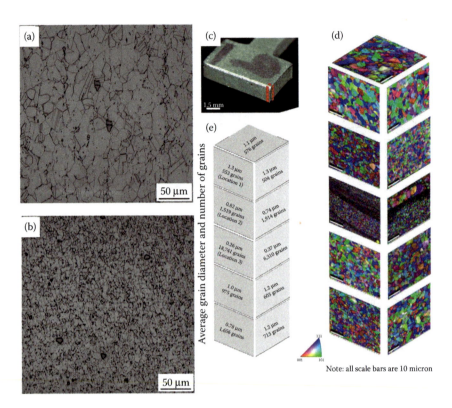

FIGURE 6.13
Microstructural images of: (a) optical microscopic (OM) images as-received IN625 filler material; (b) OM image of as-built AFS IN625, (c) sample locations along build direction of AFS IN625, (d) Euler electron back scattered diffraction (EBSD) maps showing grain size distribution along build direction of locations shown in (c), (e) number of grains and average grain sizes. (From Rivera, O. G. et al. *Materials Science and Engineering: A*, 2017. 694(Suppl. C): 1–9.) [50]

sample 2 was slightly larger than that in sample 1. This may be due to fast cooling and less grain growth after AFS in sample 2. The feed rod utilized for the study has large average grain sizes of ~15 μm and an inhomogeneous grain structure. After AFS, the average grain size measured was approximately 2.2–2.4 μm. Also, the resulting structures displayed homogeneity in tensile properties and enhanced impact resistance, which clearly surpassed the properties obtained in conventionally fabricated components. This is no less than a landmark since the issue of anisotropy is a major limiting factor in component fabrication via MAM techniques [51]. The authors concluded that resultant microstructures in AFS depend upon feeder materials, thermal cycles, and rates of cooling characteristics [47]. Also, they emphasized the need for devising a heat treatment strategy for obtaining a higher yield and tensile properties. It was concluded that the resulting components possessed wrought microstructures with mechanical and physical properties that appreciably exceeded the base metal properties [47].

Thus, from the thorough examination of experimental studies conducted on AFS, it can be concluded that AFS is a rapid friction stir-based AM process that has the capability to produce components in bulk from a wide variety of input materials. In addition to this, the deposited material exhibits fine grain structure, good metallurgical bonding, and fewer porosity defects. Lighter materials components, like Al alloys used for aircraft, special functional materials like Ni-based superalloys, and surface and bulk composites of metals, have been successfully fabricated via AFS [46,50,52,53]. Certain applications like fabricating larger structures with ultrafine granular-sized magnesium are feasible only owing to AFS [48].

6.6 Machines Utilized for Friction Stir Additive Manufacturing, Friction-Assisted Seam Welding, and Additive Friction Stir

The requirements for machines for FSAM, FASW, and AFS are similar to those for FSW, which implies that the machines utilized for FSW can be used for FSATs. There are basically two types of control systems, namely force and position control strategies. Joining/consolidation of materials during FSATs occurs under an axial forging force between consolidating materials. Material temperature plays a significant role during this process. Since the temperature is high, the material becomes softer and the axial (and forging as well) force is less and vice versa. Tool rotation, tool traverse, and axial force are very important parameters for FSATs. The prevailing temperature largely depends on the traditional parameters, that is, rotational rate and traverse rate. These two parameters are independently controlled in FSATs during the steady state. When FSATs are performed on an adapted vertical milling machine, axial force is indirectly controlled through plunge depth, whereas in professional and robotic FSW/P machines, the axial force can be independently controlled during FSATs. Advanced FSW/P systems are devised to perform operations in both force and position control modes.

In the steady state, at constant levels of rotational and traverse rate, a change in plunge depth by even a fraction of a millimeter may change the axial force by several thousand Newtons. A variation in plunge depth during the transient state may, however, result in an exceptionally high change in axial force, even three to five times greater than an equal change in the steady state. Because of this typical nature of axial force, robotic FSW/P systems must always be operated under force-controlled mode, as robotic systems are constrained by being compliant and possessing limited load capacity. Incidentally, FSAM along a complex 3D path must be performed by a robotic system, and a sound joining needs accurate position control, whereas the system is of the force-control type. Under such conditions, the axes and arms of the robotic system may undergo deflections, causing undesirable force and positional variations and making robotic FSW/P process control a challenge. In force-controlled systems, a force sensor continuously senses the force at the tool and supplies feedback to the machine's controller. When the force deviates from a preset value, the change is sensed and feedback is sent to the controller, which adjusts the position of the tool to keep the force constant. Generally, force-control mode provides greater flexibility in accommodating part material, part geometry, and tool path, as all these changes are manifested through a change in force.

The most suitable equipment can be selected on the basis of characteristics like force, sensing, accuracy, and so on, depending upon the type of joining positions, material, and application requirements. Commonly, three types of machines are utilized for FSW in the literature:

- Conventional machines capable of performing FSW
- Customized FSW machines
- Robots designed for FSW

6.6.1 Conventional Machine Capable of Performing Friction Stir Welding

Conventional milling machines can be used for FSATs since these machines include a rotating tool to machine or process the workpiece, which is the basic requirement to perform FSAT. However, FSATs demand high axial loads as compared to conventional milling machines. To fulfill these needs, the axial loads and stiffness of these machines need to be increased. These modifications can be done by enhancing their structural, decision-making, and sensing capability [54]. Such structural enhancements make the machine more robust, while the sensing capability increases the accuracy of the machine. The cost of these machines is appreciably less as compared to customized and robotic FSW machines. These machines are most suitable for producing straight-line (2D) joints for thin as well as thick materials. However, these machines fail to join material in 3D. A pictorial view of a conventional vertical milling capable of performing FSATs is shown in Figure 6.14.

6.6.2 Customized Friction Stir Welding Machines

Customized or dedicated FSW machines are designed for the highest robustness in terms of stiffness, load capacity, and so on [55]. These machines are custom-built to perform specially defined operations. They are relatively expensive as compared to conventional milling machines and are generally recommended where high flexibility (degree of freedom) is required, as in the case of component fabrication for decks of ships. Applications of these customized machines mainly include joining of high-temperature metals like titanium, steel, nickel, and so on requiring high axial loads. Other

FIGURE 6.14
Conventional retrofitted vertical milling machine capable of performing FSW/FSP/FSAM/FSAW/AFS. (From JMI, India.)

applications of customized machines are in remote areas for fabrication of large components and/or joining of small parts in large structures. However, these portable machines should be of lower weight and size, which in turn puts a load restriction on machines.

6.6.3 Robots Designed for Friction Stir Welding

The third and most advanced type of machine is robotic FSW machines. Previously, these industrial robots found limited applications owing to their restricted load capacities. However, they are being used in recent applications owing to tremendous robotic research and corresponding improvement in load capabilities. These possess the highest flexibility as compared to modified milling machines and dedicated FSW machines. They are suitable for multisided joining in a single setup. This characteristic results in reduced running cost of welding owing to reduced material handling. The biggest specific benefit of these welding robots is the feasibility of three-dimensional welding.

A comparison of these three kinds of machines in terms of cost, stiffness, flexibility, and so on is provided in Table 6.3.

6.7 Concluding Summary

FSATs, that is, FSAM, FASW, and AFS, can be a superior path toward troubleshooting of several key challenges in MAM. They mark a superior path of

TABLE 6.3

Types of FSW Machine with Their Features

Characteristics	Types of Machines			
	Milling Machine	Customized FSW Machine	Parallel Robot	Articulated Robot
Capital investment	Low	High	High	Low
Stiffness	High	High	High	Low
Flexibility	Low	Medium	High	High
Setup time	Low	High	Medium	Medium
Capacity to produce complex welds	Low	Medium	High	High

Source: Mendes, N. et al. *Materials & Design*, 2016. 90(Suppl. C): 256–265. [56]

intricate component fabrication from machining-out to building-in approaches, which is a beautiful process having its own charm. These technologies have technology level 4 (TL4) readiness and are awaiting metamorphosis from the lab to manufacturing environments. Following are the key advantages of using FSATs as compared to conventional MAM or simple fusion-based AM techniques: ability to fabricate larger components, no requirement of shielding gas and sealed chambers, freedom from surface contaminants, wide range of raw material availability, energy efficiency, no loss of alloying ingredients, reproducibility, ability to produce tailor-made microstructures, better structural capabilities, and so on. Thus, it can be concluded that FSATs have emerged as an effective way to improve the structural performance of simple geometrical structures. These techniques have broadened the alloy space and provided enhanced ease in multimaterial component manufacturing. These techniques can result in superior and high-performance alloys by virtue of imparting extremely fine and uniform microstructures.

References

1. Palanivel, S., Nelaturu, P., Glass, B., Mishra, R. S. Friction stir additive manufacturing for high structural performance through microstructural control in an Mg based WE43 alloy. *Materials & Design (1980–2015)*, 2015. 65: 934–952.
2. Rodelas, J., Lippold, J. Characterization of engineered nickel-base alloy surface layers produced by additive friction stir processing. *Metallography, Microstructure, and Analysis*, 2013. 2(1): 1–12.
3. Kalvala, P. R., Akram, J., Tshibind, A. I., Jurovitzki, A. L., Misra, M., Sarma, B. Friction spot welding and friction seam welding, 2014, Google Patents.
4. Thomas, W. M., Nicholas, E. D., Needham, J. C., Nurch, M. G., Temple-Smith, P., Dawes, C. Friction stir butt welding, 1991, G.B., USA.

5. Mishra, R. S., Mahoney, M. W. *Friction Stir Welding and Processing* 2007, ASM International, USA.
6. Siddiquee, A. N., Pandey, S. Experimental investigation on deformation and wear of WC tool during friction stir welding (FSW) of stainless steel. *The International Journal of Advanced Manufacturing Technology*, 2014. 73(1): 479–486.
7. Mishra, R. S., Ma, Z. Y. Friction stir welding and processing. *Materials Science and Engineering: R: Reports*, 2005. 50(1–2): 1–78.
8. Rathee, S., Maheshwari, S., Siddiquee, A. N., Srivastava, M. Analysis of microstructural changes in enhancement of surface properties in sheet forming of Al alloys via friction stir processing. *Materials Today: Proceedings*, 2017. 4(2, Part A): 452–458.
9. Rathee, S., Maheshwari, S., Siddiquee, A. N. Issues and strategies in composite fabrication via friction stir processing: A review. *Materials and Manufacturing Processes*, 2018. 33(3): 239–261.
10. Rathee, S., Maheshwari, S., Siddiquee, A. N., Srivastava, M. A review of recent progress in solid state fabrication of composites and functionally graded systems via friction stir processing. *Critical Reviews in Solid State and Materials Sciences*, 2017, https://doi.org/10.1080/10408436.2017.1358146; 1–33.
11. Lequeu, P. H., Muzzolini, R., Ehrstrom, J. C., Bron, F., Maziarz, R. High performance friction stir welded structures using advanced alloys, in *Aeromat Conference* 2006, Seattle, WA.
12. Baumann, J. A. *Technical Report on: Production of Energy Efficient Preform Structures* 2012, The Boeing Company: Huntington Beach, CA.
13. Yuqing, M. L., Ke, C., Huang, F., Liu, Q. L. Formation characteristic, microstructure, and mechanical performances of aluminum-based components by friction stir additive manufacturing. *The International Journal of Advanced Manufacturing Technology*, 2016. 83(9): 1637–1647.
14. Palanivel, S., Mishra, R. S. Building without melting: A short review of friction-based additive manufacturing techniques. *International Journal of Additive and Subtractive Materials Manufacturing*, 2017. 1(1): 82–103.
15. Palanivel, S., Sidhar, H., Mishra, R. S. Friction stir additive manufacturing: Route to high structural performance. *JOM*, 2015. 67(3): 616–621.
16. Rathee, S., Maheshwari, S., Siddiquee, A. N., Srivastava, M. Effect of tool plunge depth on reinforcement particles distribution in surface composite fabrication via friction stir processing. *Defence Technology*, 2017. 13(2): 86–91.
17. Zhang, Y. N. et al. Review of tools for friction stir welding and processing. *Canadian Metallurgical Quarterly*, 2012. 51(3): 250–261.
18. Su, J. Q. et al. Microstructural investigation of friction stir welded 7050-T651 aluminium. *Acta Materialia*, 2003. 51(3): 713–729.
19. Gerlich, A., Avramovic-Cingara, G., North, T. H. Stir zone microstructure and strain rate during Al 7075-T6 friction stir spot welding. *Metallurgical and Materials Transactions A*, 2006. 37(9): 2773–2786.
20. Jata, K. V., Semiatin, S. L. Continuous dynamic recrystallization during friction stir welding of high strength aluminum alloys. *Scripta Materialia*, 2000. 43(8): 743–749.
21. Xie, G. M., Ma, Z. Y., Geng, L. Development of a fine-grained microstructure and the properties of a nugget zone in friction stir welded pure copper. *Scripta Materialia*, 2007. 57(2): 73–76.

22. Barenji, R. V. Influence of heat input conditions on microstructure evolution and mechanical properties of friction stir welded pure copper joints. *Transactions of the Indian Institute of Metals*, 2016. 69(5): 1077–1085.
23. Chang, C. I., Lee, C. J., Huang, J. C. Relationship between grain size and Zener-Holloman parameter during friction stir processing in AZ31 Mg alloys. *Scripta Materialia*, 2004. 51(6): 509–514.
24. Chowdhury, S. M., Chen, D. L., Bhole, S. D., Cao, X., Powidajko, E., Weckman, D. C., Zhou, Y. Tensile properties and strain-hardening behavior of double-sided arc welded and friction stir welded AZ31B magnesium alloy. *Materials Science and Engineering: A*, 2010. 527(12): 2951–2961.
25. Chowdhury, S. M., Chen, D. L., Bhole, S. D., Cao, X. Tensile properties of a friction stir welded magnesium alloy: Effect of pin tool thread orientation and weld pitch. *Materials Science and Engineering: A*, 2010. 527(21): 6064–6075.
26. Chung, Y. D., Fujii, H., Ueji, R., Tsuji, N. Friction stir welding of high carbon steel with excellent toughness and ductility. *Scripta Materialia*, 2010. 63(2): 223–226.
27. Fernandez, J. R., Ramirez A. J. Microstructural evolution during friction stir welding of mild steel and Ni-based alloy 625. *Metallurgical and Materials Transactions A*, 2017. 48(3): 1092–1102.
28. Sato, Y. S., Urata, M., Kokawa, H., Ikeda, K. Hall–Petch relationship in friction stir welds of equal channel angular-pressed aluminium alloys. *Materials Science and Engineering: A*, 2003. 354(1): 298–305.
29. Rathee, S., Maheshwari, S., Siddiquee, A. N., Srivastava, M. Investigating effects of groove dimensions on microstructure and mechanical properties of AA6063/SiC surface composites produced by friction stir processing. *Transactions of the Indian Institute of Metals*, 2017. 70(3): 809–816.
30. Yin, Y. H., Sun, N., North, T. H., Hu, S. S. Hook formation and mechanical properties in AZ31 friction stir spot welds. *Journal of Materials Processing Technology*, 2010. 210(14): 2062–2070.
31. Buffa, G. et al. Friction stir welding of lap joints: Influence of process parameters on the metallurgical and mechanical properties. *Materials Science and Engineering: A*, 2009. 519(1): 19–26.
32. Shirazi, H., Kheirandish, Sh., Safarkhanian, M. A. Effect of process parameters on the macrostructure and defect formation in friction stir lap welding of AA5456 aluminum alloy. *Measurement*, 2015. 76(Suppl. C): 62–69.
33. Fersini, D., Pirondi, A. Analysis and modelling of fatigue failure of friction stir welded aluminum alloy single-lap joints. *Engineering Fracture Mechanics*, 2008. 75(3): 790–803.
34. Cao, X., Jahazi, M. Effect of tool rotational speed and probe length on lap joint quality of a friction stir welded magnesium alloy. *Materials & Design*, 2011. 32(1): 1–11.
35. Mehta, K. P., Badheka V. J. A review on dissimilar friction stir welding of copper to aluminum: Process, properties, and variants. *Materials and Manufacturing Processes*, 2016. 31(3): 233–254.
36. Tutunchilar, S., Haghpanahi, M., Besharati Givi, M. K., Asadi, P., Bahemmat, P. Simulation of material flow in friction stir processing of a cast Al–Si alloy. *Materials & Design*, 2012. 40(0): 415–426.
37. Chen, H.-B. et al. The investigation of typical welding defects for 5456 aluminum alloy friction stir welds. *Materials Science and Engineering: A*, 2006. 433(1–2): 64–69.

38. Rathee, S., Maheshwari, S., Noor Siddiquee, A., Srivastava, M. Distribution of reinforcement particles in surface composite fabrication via friction stir processing: Suitable strategy. *Materials and Manufacturing Processes*, 2018. 33(3): 262–269.
39. Kim, Y. G. et al. Three defect types in friction stir welding of aluminum die casting alloy. *Materials Science and Engineering: A*, 2006. 415(1–2): 250–254.
40. Crawford, R. et al. Experimental defect analysis and force prediction simulation of high weld pitch friction stir welding. *Science and Technology of Welding and Joining*, 2006. 11(6): 657–665.
41. Dehghani, K., Mazinani, M. Forming nanocrystalline surface layers in copper using friction stir processing. *Materials and Manufacturing Processes*, 2011. 26(7): 922–925.
42. Mehta, K. P., Badheka, V. J. Effects of tool pin design on formation of defects in dissimilar friction stir welding. *Procedia Technology*, 2016. 23: 513–518.
43. Trueba Jr, L. et al. Effect of tool shoulder features on defects and tensile properties of friction stir welded aluminum 6061-T6. *Journal of Materials Processing Technology*, 2015. 219: 271–277.
44. Chen, H.-B., Yan, K., Lin, T., Chen, S.-B., Jiang, C.-Y., Zhao, Y. The investigation of typical welding defects for 5456 aluminum alloy friction stir welds. *Materials Science and Engineering: A*, 2006. 433(1): 64–69.
45. Kalvala, P. R., Akram, J., Misra, M. Friction assisted solid state lap seam welding and additive manufacturing method. *Defence Technology*, 2016. 12(1): 16–24.
46. Kandasamy, K., Renaghan, L. E., Calvert, J. R., Schultz, J. P. Additive friction stir deposition of WE43 and AZ91 magnesium alloys: Microstructural and mechanical characterization, in *International Conference, Powder Metallurgy & Particulate Materials* 2013, Advances in Powder Metallurgy and Particulate Materials: Chicago, IL.
47. Calvert, J. R. Microstructure and mechanical properties of WE43 alloy produced via additive friction stir technology, in *Materials Science and Engineering* 2015, Virginia Polytechnic Institute and State University, Virginia Tech.
48. Su, J. Additive Friction Stir Deposition of Aluminum Alloys and Functionally Graded Structures, Phase I Project SBIR/STTR Programs | Space Technology Mission Directorate (STMD), 2013: US.
49. Nanci, H., Kumar, K., Jianqing, S., Dietrich, L., James, D. Additive friction stir deposition of Mg alloys using powder filler materials, in *TMS Annual Meeting & Exhibition*, February 14–18, 2016: Nashville, TN.
50. Rivera, O. G. et al. Microstructures and mechanical behavior of Inconel 625 fabricated by solid-state additive manufacturing. *Materials Science and Engineering: A*, 2017. 694(Suppl. C): 1–9.
51. Sood, A. K., Ohdar, R. K., Mahapatra, S. S. Parametric appraisal of mechanical property of fused deposition modelling processed parts. *Materials & Design*, 2010. 31(1): 287–295.
52. Kandasamy, K., Renaghan, L., Calvert, J., Creehan, K., Schultz, J. Solid-state additive manufacturing of aluminum and magnesium alloys, in *Materials Science & Technology Conference and Exhibition 2013: (MS&T'13)* 2013, Materials Science and Technology -Association for Iron & Steel Technology: Montreal, Quebec, Canada.
53. Rollie, C. J. Microstructure and mechanical properties of WE43 alloy produced via additive friction stir technology, in *Material Science and Engineering* 2015, Virginia Polytechnic Institute: Blacksburg, VA.

54. Yavuz, H. Function-oriented design of a friction stir welding robot. *Journal of Intelligent Manufacturing*, 2004. 15(6): 761–775.
55. Okawa, Y. et al. Development of 5-Axis friction stir welding system, in *2006 SICE-ICASE International Joint Conference*, Korea, October 18–21, 2006, 1266–1269.
56. Mendes, N., Neto, P., Loureiro, A., Moreira, A. P. Machines and control systems for friction stir welding: A review. *Materials & Design*, 2016. 90(Suppl. C): 256–265.

7

Applications and Challenges of Friction Based Additive Manufacturing Technologies

7.1 Introduction

FATs are friction welding–based novel manufacturing approaches. Ever since the inception of their parent technology, that is, FW, the sheer pace of evolution of variants in the form of FATs and their adoption by the engineering sector is astonishing. The imminent potential of FATs lies in the essence of their abilities to process exotic materials, for example, the buildup of hard to process alloys (Al 7075 and other 7xxx alloys) without the loss of mechanical properties, development of rib-stiffened plates (Al 2139, 2195, and other 2xxx alloys), direct deposition of stringers over fuselage skin, and so on, are some high-end, sophisticated applications. Unfortunately, nearly the entire body of information on the professional use of FATs is proprietary in nature, and this is the major hurdle in progress of the process, which otherwise is full of possibilities. These processes have added new dimensions to additive manufacturing techniques, which otherwise mainly remained limited to fusion-based processes. Although sooner or later the technologies of these processes will come out of the restrictions of proprietary holding, it is in the interest of the process itself to become mature as early as possible.

These techniques have been reported and forecasted as niche technologies in various critical sectors, including aerospace, automotive, biomedical, tooling industry, and so on. These technologies have the ability to considerably widen the alloy space, develop functionally graded materials, and fabricate components of high structural integrity and efficiency in addition to the ability to reduce many solidification-related defects. Issues like hot cracking of high-strength aluminum alloys can be eliminated via the FAT route, which in turn removes the restraint on applicability of aluminum in high-end structural applications like the aerospace and aircraft industries. A lot of fuel saving and expansion of capacities come as a benefit of this particular application. Lightweight magnesium alloys, high-temperature titanium alloys, special structural steels, graded materials, multimaterial components, and so on are some instances of high-end materials being explored. Entirely new avenues of microstructural engineering via MAM techniques have come into existence owing to these technologies. An overview of current and potential applications of different FATs is presented in this chapter. It also aims to discuss the different underlying issues in the full-fledged use of these techniques with the benefits of sustainable and green manufacturing.

7.2 Applications of Friction Based Additive Manufacturing Technologies

AM technology is a fast-expanding frontier in the manufacturing domain across industrial sectors, subsuming conventional methods. Typically, the majority of AM technologies are fusion based, such as laser and electron beam–based melting processes. Characteristically, fusion-based processes inherently suffer from limitations such as low productivity, higher manufacturing cost, issues with mechanical properties, low deposition volume, and so on. However, potential user industries such as aircraft, aerospace, nuclear and fossil energy, and transportation sectors deal with giant parts and thus require higher deposition volumes and productivities and of course guaranteed superior mechanical properties. The newer FATs fortunately suitably address almost all problems of fusion-based AMs. These techniques have the unique capability to develop multimaterial builds/components without the formation of any deleterious phases. Detailed applications of various FATs based upon their underlying strategy are covered in subsequent sections.

a. *Applications of friction joining–based FATs*: Friction joining–based AM methods are generally classed under the broad category of friction welding. This can be utilized to weld a large variety of parts with varied spatial alignment with respect to each other. The quality of welds obtained is of quite superior quality. Dissimilar welds can be easily obtained using friction welding techniques. Also, a large range of dimensions can be welded using these techniques. As presented in Figure 7.1, varied weld alignments can be utilized during friction welding.

Interestingly, these processes infuse a lower amount of energy, as the material is not melted, yet a significant part of the supplied energy remains absorbed and enhances mechanical properties through the actions of plastic deformation and dynamic recrystallization. Unlike fusion-based processes, FW methods are highly affable to

Applications and Challenges of FATs

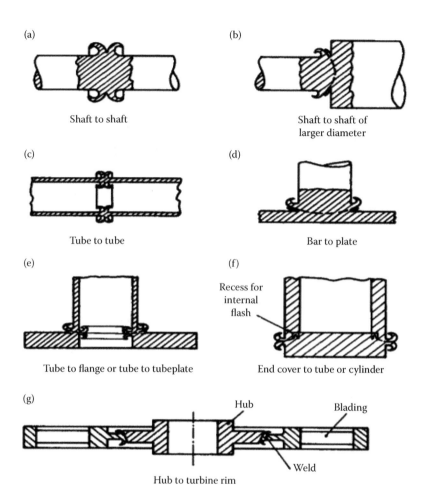

FIGURE 7.1
Configurations that can be welded via FW. (Adapted from Crossland, B. *Contemporary Physics*, 1971. 12(6): 559–574.)

dissimilar material joining. This brings about interesting benefits in the form of reduction in number of parts; cost reduction; time savings in alignment, clamping and parts assembly; and so on. For example, if a high-cost functional material is cladded over a low-cost substrate, then the composite still performs the same function, but at a substantially lower cost. Further, these processes are marked by the advantage of improved rates of material addition, which significantly reduces the cost and enhances productivity. Fortunately, most parts can be defined/classified based on round and straight/flat-feature FATs. Both these features can be developed by using RFW- or LFW-based processes, respectively.

The specific applications of conventional RFW can be summarized as under [1]:

1. Can be utilized to join two shafts (similar and dissimilar cross-sections), shaft–flange, two tubes, tube–flange, and so on
2. Welding shanks with drills of entirely dissimilar constitution
3. Valve heads for internal combustion (IC) engines to their stems
4. Pipes of various composition to the flanged supports
5. Bearing housing to automobile casing
6. Individual gears and their clusters in a gear box
7. Aluminum bars to steel pieces, which is a very interesting and useful application.

With the use of RFW as an AM tool, it can be successfully utilized in the fabrication of honeycomb structures, multimaterial dies (forging), multimaterial rolls (metal forming), and fully enclosed (cavities/feature) components [2].

The main applications of conventional LFW are as follows:

1. Manufacturing of compressor BLISKs for military/civilian gas turbines/aircraft engines, which have a niche market. The BLISK thus formed results in approximately 30% weight savings in the aerodynamic industry [3].
2. Repair of BLISKS [3]
3. Huge-sized components for the aircraft and space industries
4. Huge-sized components for the oil, gas, and energy industries
5. Plastic pipe joints
6. Welding of various automotive components
7. Welding a wide domain of metals like steel, titanium, nickel, and so on and the corresponding metal-based alloys
8. Giants like Rolls-Royce, Pratt and Whitney, and so on utilize LFW to produce and repair disc, compressors, turbine engines, and so on
9. Joints in wheel rims in various parts of transport systems
10. Welding of braking blocks
11. Welding parts fabricated from thermally hardened plastic
12. Fabricating flat-link hinge chains, gears, electrical buses

b. *Applications of friction deposition–based FATs*: Friction deposition–based FATs such as FD and FS techniques utilize the hot forging action of the consumable rod that refines the microstructure of the

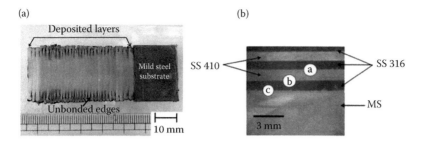

FIGURE 7.2
Typical macrographs of: (a) friction deposition (Adapted from Dilip, J. J. S. et al. *Transactions of the Indian Institute of Metals*, 2011. 64(1): 27.) [5], (b) friction surfacing, showing multimaterial build. (Adapted from Dilip, J. J. S. et al. *Materials and Manufacturing Processes*, 2013. 28(2): 189–194.) [4]

deposit/coatings. Existing research has described the development of components from various types of metal/alloys like aluminum, steel, and so on, as well as composites. Fabricated components from both of these techniques are free from porosity or slag inclusions, fumes, and radiation and possess refined microstructures similar to most of the other friction welding–based variants. FS is fundamentally a surface engineering technique that is gaining huge predominance in the current manufacturing arena. It can be successfully utilized for various geometrical positions. The extension of conventional FS processes to AM techniques may lead to the development of rib-on-plate structures, components with fully enclosed internal features, and so on [4]. Similar to FS, FD-based AM methods can be successfully utilized to develop 3D components of different types of materials. Typical builds fabricated via FD and FS are shown in Figure 7.2a and b, respectively.

c. *Applications of friction stir–based FATs*: These are the classes of processes based on principles of friction stir welding. As discussed in detail in Chapters 3 and 6, FSAM, FASW, and AFS are the three friction stir joining–based FATs. A special characteristic of these processes is that they are accompanied by severe plastic deformation and simultaneous dynamic recrystallization. This leads to wrought microstructures and properties comparable or superior to those of the parent metals. Additionally, these can be utilized to obtain tailored microstructures with graded properties easily. Control of microstructures is quite easy in these processes. A major limitation includes the fact that machining is required after each layer, which makes the process expensive and time consuming except in the case of AFS. Also, tooling needs to be designed very carefully, as the design of the tool affects the material flow and properties of the fabricated components. The fixturing system needs special attention, mainly for components with large z-heights.

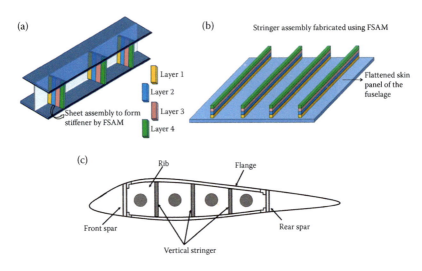

FIGURE 7.3
Stiffener assembly via FSAM process for aerospace industry; (a) I-beam with transverse stiffener, (b) stringers over flattened skin panel of fuselage, (c) air foil C-S that depicts integration in fabricating ribs and stringers in wing spar web. (Adapted from Palanivel, S. et al. *JOM*, 2015. 67(3): 616–621.) [6]

Following are the specific applications of FSAM in fabrication of:

1. High-performance alloys
2. Builds of full consolidation, high yield, and tensile strength
3. Builds of high structural performance
4. Energy-efficient builds
5. Creep-resisting structures
6. Future prospects in the aircraft, automotive, fossil, and nuclear sectors, especially stringer assemblies, as shown in Figure 7.3.
7. Homogeneous bulk composite, functionally graded structures, sandwich structures, and so on, as illustrated in Figure 7.4.

Following are the specific applications of FASW [7]:

1. Multilayer/multimaterial seams of high- and low-temperature materials
2. A variety of industrial welding applications in aircraft, marine, nuclear, and processing industries, and so on
3. Defect-free seam-welded components
4. Can be utilized for different alloy combinations
5. Can be utilized in development of leak-proof joints, for example, fuel tanks, mufflers, and so on

Applications and Challenges of FATs

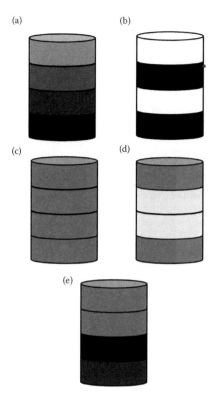

FIGURE 7.4
Illustration of some possible combinations that can be fabricated using FSAM (a) fully gradient, (b) alternate gradient, (c) bulk composite, (d) sandwich structure, (e) tailored composite.

6. Can be used for friction cladding. Its multitrack seam cladding can be used in next-generation nuclear plants
7. Promising potential for additive manufacturing, especially for materials and material combinations that are so far not compatible with the prevailing AM technologies

AFS has a distinct merit over FATs wherein localized deposition is preferred either about a rotational (RFW) or linear (LFW) symmetry. In the case of AFS, there is full special flexibility for the laying-up of the build, as the path of the deposit can be controlled free from 3D space.

Following are the specific applications of AFS:

1. Can be successfully utilized for fabrication of fully dense, economical components possessing wrought microstructures from a considerable range of metal and alloys in bulk quantities
2. Can easily fabricate intricate structures without the application of a machining step

TABLE 7.1
Current and Potential Applications of FATs

Process	Present Applications	Probable Applications
Rotary friction welding [2]	Two shafts (similar and dissimilar cross-sections), shaft–flange, two tubes, tube–flanges, etc. in tools, valves, automotive components	Multimaterial components like metal-forming roll, components with fully enclosed cavities, honeycomb structures, etc.
Linear friction welding [8]	Manufacturing and repair of aircraft BLISKs, automobile braking blocks, large-sized pipes and other components, etc.	Gradient structure fabrication at large scale, airframe brackets, etc.
Friction deposition [9]	Development of ferrous and nonferrous metal deposits, metal–metal composites, etc.	Injection moulding dies and tooling, parts with fully enclosed cavities, circular parts with multimaterial composition
Friction surfacing [4]	Coatings for high wear and corrosion applications, repair applications	3D multimaterial, aircraft frames such as rib-on-plate structures, repairing of dies, etc.
Friction stir additive manufacturing [10,11]	Producing structural components from Al, Mg alloys, preform fabrication	Functionally gradient materials, stringers in aerospace fuselage, etc.
Friction-assisted (lap) seam welding [12]	A variety of industrial welding applications in automotive, aircraft, marine, nuclear, and processing industries, etc.	Cladding and AM of similar as well as dissimilar materials, producing high-strength seam welds, AM of wide range of materials, etc.
Additive friction stir [13–16]	Fabrication of high-strength ultrafine-grained magnesium alloys	Coating of shaft journals, airframe structures like bulkheads and stiffeners, etc.

The process capabilities and potential have resulted in its commercialization and industrial adoption, specifically employed for laying-up of stiffeners over ribs, stringers to body frame connections and stabilizer parts, and so on. Further, giant structures for the nuclear and fossil energy industries (e.g., strengthening of flat and hollow surfaces with stiffeners) are typical examples of applicability of this process. The current and probable applications of each of FAT are summarized in Table 7.1.

7.3 Challenges of Friction Based Additive Manufacturing Technologies

Friction based additive techniques involve utilization of either friction welding setups or friction stir welding setups, along with a CNC machine as a machining tool. Robots (in place of conventional machine setups) are also in use, but are very limited. To date, these individual setups are exclusively used in completing 3D fabrication of parts. A significant challenge involved is fabrication of such modelers/equipment in which both type of operations, that is, additive joining and machining, can be performed at a single platform. For this purpose, design engineers, researchers, and manufacturers from the fields of both additive manufacturing and friction based processes need to come together to design such equipment.

Another main challenge can be described in terms of nonhomogeneous microstructures over particular layers. As in most friction welding (like friction stir welding)–based processes, there are mainly four zones, namely the weld/nugget zone, thermomechanical-affected zone, heat-affected zone, and parent metal zone. Variation in microstructural features takes place in these different zones, which also affects their longitudinal strength. However, this limitation can be utilized in the fabrication of graded structures by careful monitoring and control.

7.4 Conclusion

Despite tremendous innovations in these areas, a great deal of scope of new research and development still remains. Consistent efforts of researchers based upon improved knowledge of complexities involved in these processes can carve paths of complete commercial success of these technologies. In the current technical scenario, FATs require intensive investigations before these processes are well accepted commercially by the industry. Thus, much work and serious research needs to be accomplished in order to transform (FATs) from a meager weapon in the armor of manufacturing engineers to a full-scale technology and to fully utilize its potential.

References

1. Crossland, B. Friction welding. *Contemporary Physics*, 1971. 12(6): 559–574.
2. Dilip, J. J. S., Janaki Ram, G. D., Stucker, B. E. Additive manufacturing with friction welding and friction deposition processes. *International Journal of Rapid Manufacturing*, 2012. 3(1): 56–69.
3. Turner, R., Gebelin, J. C., Ward, R. M., Reed, R. C. Linear friction welding of Ti–6Al–4V: Modelling and validation. *Acta Materialia*, 2011. 59(10): 3792–3803.
4. Dilip, J. J. S., Babu, S., Rajan, S.V., Rafi, K. H., Janaki Ram, G. D., Stucker, B. E. Use of friction surfacing for additive manufacturing. *Materials and Manufacturing Processes*, 2013. 28(2): 189–194.

5. Dilip, J. J. S., Kalid, R. H., Janaki Ram, G. D. J. A new additive manufacturing process based on friction deposition. *Transactions of the Indian Institute of Metals*, 2011. 64(1): 27.
6. Palanivel, S., Sidhar, H., Mishra, R. S. Friction stir additive manufacturing: Route to high structural performance. *JOM*, 2015. 67(3): 616–621.
7. Kalvala, P. R., Akram, J., Tshibind, A. I., Jurovitzki, A. L., Misra, M., Sarma, B. Friction spot welding and friction seam welding, 2014, Google Patents.
8. Slattery, K. T., Young, K. A. Structural assemblies and preforms therefor formed by friction welding, 2008, Google Patents.
9. Dilip, J. J. S., Janaki Ram, G. D. Microstructures and properties of friction freeform fabricated borated stainless steel. *Journal of Materials Engineering and Performance*, 2013. 22(10): 3034–3042.
10. Yuqing, M. L., Ke, C., Huang, F., Liu, Q. L. Formation characteristic, microstructure, and mechanical performances of aluminum-based components by friction stir additive manufacturing. *The International Journal of Advanced Manufacturing Technology*, 2016. 83(9): 1637–1647.
11. Palanivel, S., Nelaturu, P., Glass, B., Mishra, R. S. Friction stir additive manufacturing for high structural performance through microstructural control in an Mg based WE43 alloy. *Materials & Design (1980–2015)*, 2015. 65: 934–952.
12. Kalvala, P. R., Akram, J., Misra, M. Friction assisted solid state lap seam welding and additive manufacturing method. *Defence Technology*, 2016. 12(1): 16–24.
13. Rodelas, J., Lippold, J. Characterization of engineered nickel-base alloy surface layers produced by additive friction stir processing. *Metallography, Microstructure, and Analysis*, 2013. 2(1): 1–12.
14. Rivera, O. G. et al. Microstructures and mechanical behavior of Inconel 625 fabricated by solid-state additive manufacturing. *Materials Science and Engineering: A*, 2017. 694(Supplement C): 1–9.
15. Calvert, J. R. Microstructure and Mechanical Properties of WE43 Alloy Produced Via Additive Friction Stir Technology, in *Materials Science and Engineering* 2015, Virginia Polytechnic Institute and State University: Virginia Tech.
16. Su, J. *Additive Friction Stir Deposition of Aluminum Alloys and Functionally Graded Structures, Phase I Project SBIR/STTR Programs | Space Technology Mission Directorate (STMD)* NASA report. 2013. US.

8
Conclusion

8.1 Introduction

Friction based additive techniques are game-changing processes that have the potential to significantly expand the boundaries of metal additive manufacturing (MAM). Each sector requiring applications of structurally sound, ergonomically intricate, and functionally simple components within stipulated time requirements is a candidate for these FATs. These mainly include the biomedical, aircraft, space, defense, marine, transportation, and architecture sectors. Every process belonging to FAT group has specific benefits and limitations. Since some of these processes are in their infancy, there exists considerable scope to develop them to become effective alternatives as a professionally and industrially acceptable manufacturing philosophy. A discussion on these has been included at the end of every relevant chapter. However, some points that need to be specially highlighted are included in the subsequent section.

8.2 Concluding Summary

FATs can be a superior path toward troubleshooting of several key challenges in MAM. They mark an effective manufacturing technique for intricate component fabrication from machining-out to building-in approaches, which is a magnificent process having its own charm. These technologies have technology level 4 readiness and are awaiting metamorphosis from the lab to manufacturing environments. The following are the key advantages of using FATs as compared to conventional MAM techniques: ability to fabricate larger components, no requirement of shielding gas, no requirement of vacuum and sealed chambers, freedom from surface contaminants, wide range of raw material suitability, energy efficiency, no loss of alloying ingredients, reproducibility, ability to produce tailor-made microstructures, better structural capabilities, and so on.

The key conclusions of this book can be listed as:

- Advancements and the corresponding limitations in MAM have led to a rigorous quest among researchers to look for solutions.
- Solid-state FATs have come across as landmark solutions to problems encountered in full-scale utilization of MAM.
- Size independence, homogeneous microstructures, structural strength, shape flexibility, easiness of process, and so on are the key advantages of FATs.
- There are around 20 friction based welding and processing techniques, as discussed in Chapter 3.
- Out of these 20 variants of friction welding, 7 are well poised to be developed in the category of FATs. These are: linear friction welding (LFW), rotary friction welding (RFW), friction deposition (FD), friction surfacing (FS), friction stir additive manufacturing (FSAM), friction-assisted seam welding (FASW), and additive friction stir (AFS).
- FATs have emerged as an effective way to improve the structural performance of simple geometrical structures. However, these can be successfully extended to complicated shapes.
- FATs have broadened the alloy space and provided enhanced ease in multimaterial component manufacturing.
- In short, FATs are unique techniques that can theoretically address the majority of MAM issues and can thus probably have massive utilization. However, exhaustive cost-benefit and economic analysis are required to transform these techniques into full-fledged commercial technologies.
- FATs eliminate the challenge of liquid–solid transformations, poor microstructures, and structural strength issues encountered in fusion-based MAM and result in highly desirable wrought microstructures.
- LFW and RFW can be classed as friction joining–based FATs.
- FD and FS can be classed as friction deposition–based FATs.
- FSAM, FASW, and AFS can be classed as friction stir–based FATs.
- RFW is the oldest member of the FW family.
- RFW offers self-cleaning capabilities, high rectitude, fast and reliable processes, minimized errors, and freedom from shielding gases.
- RFW cannot be utilized for thin-walled structures.
- RFW is especially suitable for welding dissimilar materials with overmatched tensile characteristics to a base material.
- LFW results in elimination of mechanical joints, stamping, and heavy-duty finishing. It has been used in repair and is an established process for economic engine manufacture.

- LFW setup is quite expensive and complex, which is a major process limitation.
- BLISK fabrication and repair is the single biggest application of LFW.
- FD results in good bonding strength; higher deposited thickness; fine-grained microstructures; development of functionally graded, multimaterial components; and homogeneously distributed composites.
- A major limitation of FD is the requirement of machining each layer, which increases cost and time requirements.
- Chief potential applications of FD include fabrication of metal–metal composites, which are difficult to fabricate by conventional AM techniques.
- FS results in sound deposits in similar as well as dissimilar materials.
- A major limitation of FS is machine dependence, especially in the case of the deposition area.
- Chief potential applications of FS are fabrication of components with enclosed cavities, functionally graded materials, and so on.
- FSAM can result in superior and high-performance alloys by virtue of imparting extremely fine and uniform microstructures.
- Chief issues in components fabricated via FSAM are: remaining joint lines, cavities, crack, band formation, hook, and kissing bond defects, which can be eliminated by proper selection of process parameters.
- Chief applications of FSAM include stiffener assembly, fuselage, stringers for aircraft, low-weight composites, alloys for structural performance, and so on.
- FASW has recently been introduced and is slightly unexplored owing to the meager amount of research reported till date.
- A major limitation is the need to machine after each layer, which amounts to an increase in machine and manufacturing costs.
- FASW can be utilized for a wide variety of similar and dissimilar material combinations. Probable applications include multilayer and multiseam welded components in sectors like aircraft, the nuclear industry, fuel tanks, and so on.
- AFS is a complete AM technology and fully adheres to the principle of layer-by-layer fabrication rather than consolidation, as in other FATs.
- AFS effectively deals with the issue of anisotropy of the AM process and is one of the only core AM processes where lateral strength is comparable to/greater than longitudinal strength.
- A major limitation of AFS is the requirement of a specially designed tooling system.

- AFS can be successfully utilized for fabrication of fully dense, economical components possessing wrought microstructures from a considerable range of metals/alloys in bulk quantities.

Despite tremendous innovations in these areas, a significant scope of new research and development still remains. In the current technical arena, FATs require intensive investigations before these processes are well accepted commercially by the industry. A consistent effort by researchers based upon improved knowledge of the complexities involved in these processes can carve paths of complete commercial success for these technologies.

8.3 Future Scope

Future success in the aforementioned areas mainly depends upon the ability to manufacture high-potential, robust, and complicated shape geometries with relative ease. The future prospects of this technology are extremely promising. Attainment of newer materials, reliable and robust communication interfaces, better hardware, the ability to fabricate complicated designs, no limitation on sizes, the ability to tailor material properties for strength and microstructure by control of grain and phase transformations, a wider range of alloys and composites, and so on will be the key focuses of research in the near future. This would pave paths for phase and microstructural engineering. Efforts to optimize deposition rates can considerably improve process performance. Controlling the quality of the components is also a key research area. These developments would lead to many innovative advancements in the field of FATs. These techniques will facilitate the fabrication of bulk and functionally graded composites to a great extent. Also, these FATs can be utilized in fabricating sandwich as well as honeycomb structures. Further, they can be utilized in developing functionally graded foams with similar or dissimilar facing materials. It is a matter of common knowledge that FW can be used to weld a wide range of materials. Since FATs are based upon FW, the type of materials that can be used to obtain high structural components, sandwich structures, metallic foams, composites, and functionally graded materials via FATs is very broad. In the future, components made via the FAT route may find applications in launch vehicles, intricate structural support components, webbed and ribbed frames, tools, tailored graded structures, wear resistant and anticorrosive coats, tailored microstructures, controlled grain growth, and so on.

To summarize, we can say that FATs are innovative techniques that are based on the confluence of friction welding and layer-by-layer additive manufacturing principles. However, much research and development are required, and there is a long way to go. If properly utilized, these technologies can work wonders. The authors' hard work and efforts are meager and modest with an aim to present a few pearls of knowledge to the scientific community. A lot of oysters still lie unexplored and are awaiting discovery.

Index

A

AA, *see* alumin(i)um alloys
Accuracy
   in friction welding, 71
   in metal additive manufacturing, 34
Additive friction stir (AFS), 3, 51, 52, 97, 114–119, 137–138
   applications, 117, 118–119, 131–132, 132, 138
   benefits and limitations, 54
   machines, 117–119
   microstructural characterization, 114–117
   working principle, 51, 114; *see also* friction stir additive manufacturing (FSAM)
Additive manufacturing/AM (basic references)
   advantages, 24–30
   applications, 30–32
   classification of techniques, 20–23
   data flow, 19, 20
   definition and terminology, xvii, 2
   future scope, 36–37, 138–139
   historical development, *see* historical and chronological development
   key names of techniques, 11
   limitations and challenges, 2, 3, 24–30
   in linear friction welding, 67–69
   metal-based, *see* metal-based additive manufacturing
   processes, *see* processes
   in rotary friction welding, 84
   working principles, *see* working principles
Additive manufacturing/AM techniques and technology, 11–40
   friction deposition-based, *see* friction deposition; friction surfacing
   friction joining-based, *see* joining techniques; welding
   friction stir welding-based, *see* friction stir welding
Advance/advancing side in friction stir welding, 101
Aerospace, 31
Aircraft, 31
   friction based additive manufacturing technologies, 132
   linear friction welding, 68, 128
AISI 304 stainless steel
   friction-assisted seam welding, 112, 113, 114
   friction deposition, 76, 77, 78, 79
AISI 310 stainless steel, 64
AISI 316 stainless steel, 91, 92
AISI 410 stainless steel, 90, 91, 92
Alloys
   aluminum, *see* alumin(i)um alloys
   CoCrFeNi high-entropy alloy powder, 78, 81, 82
   loss of alloying elements, 35
   magnesium-based WE53 alloys, 52, 106, 107, 115
   nonferrous metal alloy, 80–81
Alumin(i)um alloys (AA), 98, 105, 125
   AA 5083, 78, 81, 82, 108, 109
   AA 7075, 53, 106, 108, 109, 125
   Al 6061, 112, 113, 114
American Society for Testing and Materials (ASTM) F42 Committee guidelines, 23, 24
Angular friction welding, 45
Applications and needs, 4–5, 126–132; *see also specific methods*
Architectural industry, 32
Articulated robot, 120
Artistic industry, 31–32
ASTM F42 Committee guidelines, 23, 24
Automobile industry, 31
Axial pressure/forces
   friction-assisted seam welding, 111, 112, 117

141

Axial pressure/forces (*Continued*)
  friction deposition, 77
  friction stir additive manufacturing, 104, 117
  friction stir processing, 117
  friction stir welding, 101, 117
  friction surfacing, 87–88
  rotary friction welding, 63
Axis (tool) in friction stir welding, 101–102, 104–105

B

Binder jetting, 4, 24
Blade + disk (BLISK), 67, 68, 72, 128, 137
Bladed rotors, integrated (IBRs), 68
BLISK (blade + disk), 67, 68, 72, 128, 137
Bonding (poor/defective)
  friction deposition in manufacture of metal–metal composites deposition, 80
  friction stir additive manufacturing, 110
  friction surfacing, 93
Borated stainless steel, 78, 79–80
Brazing, friction, 49
Build height in friction stir additive manufacturing, 104
Bulk materials, *see* raw materials

C

Car (automobile) industry, 31
Chronological development, *see* historical and chronological development
Cladding, 131, 132
  friction co-extrusion, 48
CNC machining, 4, 84
CoCrFeNi high-entropy alloy powder, 78, 81, 82
Co-extrusion cladding, friction, 48
Computer numerical control (CNC) machining, 4, 84
Consumable rod, 1, 75
  friction deposition, 50, 75, 76, 77, 78, 81, 128
  friction stir welding, 97
  friction surfacing, 84, 85, 88

Continuous drive friction welding (CDFW), 61, 63, 72
Continuous dynamic recrystallization (CDRX), 84
Continuous seam weld, 111
Cracking, 35
Cusing, laser, 23, 29, 33
Customized friction stir welding machines, 118–119, 120

D

Delamination, 35
Deposition processes, 23
  deposition rate in metal additive manufacturing, 34; *see also* directed energy deposition; electrodischarge deposition; friction deposition; fused deposition modeling; laser metal deposition; selective layer chemical vapor deposition; shape deposition modeling; shaped metal deposition
Digital metallization, 27
Dimensional accuracy in metal additive manufacturing, 34
Direct drive (continuous drive) rotary friction welding (CDFW), 61, 63, 72
Direct metal deposition (DMD), 24, 28
Direct metal laser sintering (DMLS), 12, 18, 29, 33
Directed energy deposition, 4, 23, 24
Discontinuous dynamic recrystallization (DDRX), 84
Dynamic recrystallization (DRX)
  friction deposition, 77, 79
  friction stir additive manufacturing, 107
  friction stir welding, 105
  friction surfacing, 89
  rotary friction welding, 84

E

Easy clad, 28
Electrodischarge deposition, 29
Electron beam melting, 3, 4, 16, 29

Index

Energy source, classification of techniques based on, 22; *see also* directed energy deposition; stored energy friction welding
Extrusion
  friction, 48
  material, 24

F

Ferrous metal deposits, 79–80
Flywheel (inertia) friction welding (IFW), 42, 61, 63, 63–64, 71
Friction-assisted (lap) seam welding (FASW), 3, 46, 52, 61, 97, 110–113, 137
  benefits, 5–6, 54
  limitations/challenges, 7, 54
  machines, 117–119
  needs and applications, 4–5, 7, 130–131, 132, 137
  status of research and recent developments, 111–113
  working principle, 51, 111
Friction based additive manufacturing (basic references only)
  applications, 126–132
  concluding summary, 135–139
  definition/terminology, xvii
  hybrid, 44–52, 100
  needs and applications, 4–5, 132
Friction based additive manufacturing technologies/FATs (basic references only), 1–9
  benefits and limitations, 54
  joining, *see* joining techniques; welding
Friction brazing, 49
Friction co-extrusion cladding, 48
Friction deposition (FD), 3, 4, 7, 50, 75–83, 94, 136, 137
  applications, 128–129, 132, 137
  benefits and limitations, 54, 82–83
  friction surfacing compared with, 83
  general features and experimental results, 76–82
  working principle, 50; *see also* friction surfacing

Friction deposition-based additive manufacturing technologies, *see* friction deposition; friction surfacing
Friction extrusion, 48
Friction hydropillar processing, 48
Friction joining, *see* joining techniques; welding
Friction plunge welding, 49
Friction processing techniques vs friction joining, 46
Friction stir additive manufacturing (FSAM), 3, 4, 51, 52, 53, 97, 102–110
  applications, 129, 130, 132, 137
  benefits, 54
  friction assisted seam welding-based AM method compared with, 110
  limitations/challenges/defects, 54, 108–110, 132–133
  machines, 117–119
  timelines, 53
  working principles, 102–104; *see also* additive friction stir
Friction stir processing (FSP), 50, 100
  additive, *see* additive friction stir
Friction stir welding (FSW), 4, 97–124, 128–132
  applications, 98–100, 128–132
  general aspects, 98–102
  machines, 117–119
  processes based on, 52, 97–124
  terminology, 101–102
  working principles, 98–99, 102–104, 111, 114
Friction stud welding, 49
Friction surfacing (FS), 3, 4, 7, 45, 83–94, 136, 137
  applications, 89–91, 93–94, 128–129, 132
  benefits and limitations, 54, 93
  general features and status of research, 89–92
  process parameters, 85–88
  working principle, 45, 84
Friction taper stitch welding, 47
Friction transformation hardening, 46
Friction welding, *see* welding
Frictional heat, *see* heat

Fused deposition modeling (FDM), 12, 13, 14, 15, 16, 18, 25
Fusion-based processes, 4, 5, 6, 97
   limitations/disadvantages, 52, 55, 82–83, 93, 126
   metal-based additive manufacturing (MAM), 22
   powder bed fusion, 4, 23, 24, 33, 34
Future scope of AM, 36–37, 138–139

G

Grain size
   additive friction stir process, 114, 115, 116
   friction stir additive manufacturing, 105–107

H

Hall–Petch effect, 107
Heat (frictional), 41
   in friction stir welding, 98, 111, 114
   in rotary friction welding, 60, 61
Historical and chronological development (timelines), 12–19, 52, 53
   friction Welding, 41–42
   metal-based additive manufacturing, 33–34
Hook (defect) in friction stir additive manufacturing, 109
Hybrid friction based AM, 44–52, 100

I

Implants, biomedical, 32, 36
Inconel, 78, 115
Industrial applications and systems, 13, 20, 31–32
Inertia friction welding (IFW), 42, 61, 63, 63–64, 71
Inkjet printing, 4, 25
Integrated bladed rotors (IBRs), 68

J

Jetting
   binder, 4, 24
   material, 24
Jewelry, 31–32

Joining techniques (friction based), 7, 41–74
   in additive manufacturing, 59–74
   applications, 126–128
   classification based on, 22
   working principles, 45–51, 60–61, 65–66, 84, 98–99, 102–104, 111, 114; *see also* welding

K

Keyhole porosity, 34
Kissing bonds in friction stir additive manufacturing, 110
Klopstock's friction seam welding, 46

L

Laminated object manufacturing, 12, 14, 26, 33; *see also* sheet lamination
Lap seam welding, *see* friction-assisted (lap) seam welding
Laser-additive manufacturing (LAM), 26, 33
   laser-engineered net shaping (LENS), 24, 26, 33, 81
Laser cusing, 23, 29, 33
Laser melting, 27
   selective (SLM), 4, 12, 17, 18, 28, 29, 33, 34, 81
Laser metal deposition, 27
Laser sintering, 12, 16, 21, 22, 26
   direct metal (DMLS), 12, 18, 29, 33
   selective (SLS), 4, 12, 14, 23, 25, 26, 33, 34
Layer-by-layer method (layers deposited), 5
   friction deposition, 81, 82–83
   friction stir additive manufacturing, 100, 102, 104, 105–106, 107
   friction surfacing, 84, 85, 89, 90–91, 91
Layer thickness, 23, 83, 93
   friction stir additive manufacturing, 102
   metal-based additive manufacturing, 34
Leading edge in friction stir welding, 101

Linear friction welding (LFW), 3, 45, 65–69, 70, 70–72, 136–137
  additive manufacturing with, 67–69
  advantages, 68, 70–71
  applications, 128, 132, 136
  comparisons with RFW, 65, 68, 69, 70
  factors affecting, 66–67
  limitations, 71–72
  working principles, 45, 65–69
Liquid-phase additive manufacturing techniques, advantages over, 54
LUC's friction seam welding, 46

M

Machine(s)
  classification of technique based on, 23
  friction stir additive manufacturing, 117–119
  robotic, *see* robotic machines
Machined surfaces, friction-deposited, 76
Magnesium-based WE53 alloys, 52, 106, 107, 115
Materials
  classification of techniques based on material delivery, 23
  classification of techniques based on materials used, 22–23
  consumable rod, 17
  extrusion, 24
  in friction welding, 69, 70, 71
  jetting, 24
  joining, *see* joining; *see also* raw materials
Mechtrode, friction surfacing, 84, 85, 89, 91
Medical industry, 26, 28, 29, 32
  implants, 32, 36
Melting
  electron beam, 3, 4, 16, 29
  laser, *see* laser melting
Metal-based additive manufacturing (MAM), 2–3, 5, 33–36, 59, 135, 136
  fusion-based, 22, 82, 93, 115
  limitations/problems, 34–36, 82–83, 93

  machines, 23
  techniques, 33–36; *see also* digital metallization; direct metal deposition; direct metal laser sintering; laser metal deposition; shape deposition modeling; shaped metal deposition
Metal–metal composites, 80–82
Metallic components (fabrication), 3, 4, 5
  friction stir welding, 107
  friction surfacing method, 83
Microstructural properties
  additive friction stir, 114–117
  friction welding, 70
Milling machines, friction stir additive manufacturing, 118, 120
Motor (automobile) industry, 31
Multitrack scheme
  friction cladding, 131
  friction surfacing, 84, 88, 89
  friction-assisted seam welding, 111, 112, 113

N

Needs and applications, 4–5, 126–132; *see also specific methods*
Nonconsumable rod, 1
  friction stir welding-based techniques, 46, 51, 97, 98, 102, 110, 111, 114
Nonferrous metal alloy, 80–81
Nonmetals
  machines, 23
  raw material, 2

O

Orbital friction welding, 47

P

Parallel robot, 120
Patents, 12, 14–19, 53
  expiry, 13
Pin (tool) in friction stir welding, 101, 103, 106, 107, 109, 111

Plastic deformation
    additive friction stir, 114
    friction stir additive manufacturing techniques, 105
    friction-assisted seam welding, 111
Plunge depth in friction stir welding, 101, 104, 111, 117
Plunge welding, friction, 5
Pores and porosity
    friction stir additive manufacturing, 109–110
    metal additive manufacturing, 34
Powder bed, 23
    fusion, 4, 23, 24, 33, 34
Powder feed, 23
Processes (and process chain/steps/ parameters), 19, 25–29
    common processes, 24, 25–29
    friction stir additive manufacturing, 102–103
    friction surfacing, 85–88
    joining techniques, 43–54, 63–64
Prometal, 28

R

Radial friction welding, 47
Raw (bulk) materials
    additive friction stir process, 114
    classification of AM techniques based on (physical state), 20, 21
    nonmetal, 2
Recrystallization
    friction deposition, 77, 79
    friction stir additive manufacturing, 105, 107
    friction stir welding, 105
    friction surfacing, 89
    rotary friction welding, 84
RepRap technology, 12, 18
Retreating side in friction stir welding, 101
Robotic machines, 132–133
    friction stir welding, 117–118, 119, 120
Rods, see consumable rod; nonconsumable rod
Rotary friction welding (RFW), 3, 43, 45, 60–65, 69, 70, 70–72, 136
    additive manufacturing with, 64–65
    advantages/benefits, 42–43, 54, 70–71
    applications, 64–65, 128, 132, 136
    comparisons with LFW, 65, 68, 69, 70
    limitations, 54, 64–65, 71–72
    variants, 61, 62
    working principle, 45, 60–61
Rotation(al) speed
    friction deposition, 76–77
    friction stir welding, 101, 104, 107, 110, 112, 115
    friction surfacing, 85–87
    rotary friction welding, 63
Rough surfaces
    friction deposition on, 76
    metal additive manufacturing and surface roughness, 34

S

Seam welding, see friction-assisted (lap) seam welding (FASW)
Selective laser melting (SLM), 4, 12, 17, 18, 28, 29, 33, 34, 81
Selective laser sintering (SLS), 4, 12, 14, 23, 25, 33, 34
Selective layer chemical vapor deposition, 29
Shape deposition modeling, 21, 22, 29
Shaped metal deposition, 29
Sheet lamination, 4, 24
Shoulder (tool) in friction stir welding, 101, 104, 106, 107, 111
Single-track scheme
    friction surfacing, 89, 91
    friction-assisted seam welding, 111, 112
Sintering, see laser sintering
Sliced stereolithography, 20
Solid ground curing (SGC), 12, 14, 25
Stack metal sheets in friction stir additive manufacturing, 103
Stainless steel
    AISI 304, see AISI 304 stainless steel
    AISI 310, 64
    AISI 316, 91, 92

Index

AISI 410, 90, 91, 92
   borated, 78, 79–80
Steel, *see* stainless steel
Stereolithography (SLA), 12, 14, 15, 16, 25, 26, 32
   sliced, 20
Stir processes, *see* additive friction stir; friction stir additive manufacturing
Stored energy (inertia) friction welding (IFW), 42, 61, 63, 63–64, 71
Strength and strengthening (of fabricated components)
   friction-assisted seam welding, 113–114
   friction stir additive manufacturing, 107–108
Stud welding, friction, 49
Surfaces, *see* machined surfaces; rough surfaces
Surfacing, *see* friction surfacing

T

Tapered stud (friction taper stitch welding), 47
Temperature
   additive friction stir, 115
   friction deposition, 77
Tensile strength
   friction-assisted seam welding, 113, 114
   friction stir additive manufacturing, 107, 108, 110
Third body friction welding, 49
Three-dimensional (3D) fabrication/printing, 1, 12, 133
   friction deposition, 75, 76, 77, 79, 80, 94
   friction joining, 52, 59, 72
   friction stir welding, 111, 113, 117, 118, 119, 129, 131
   friction surfacing, 84, 89, 90, 91, 94
   historical development, 12, 14, 15, 16, 17, 18, 19
   metal additive manufacturing techniques, 33
Tilt angle in friction stir welding, 101, 109–110, 114

Timelines, *see* historical and chronological development
Titanium powder-filled AA 5083-H112, 78, 81
Tool in friction stir welding
   terminology, 101–102
   variables, 104–105
Trailing edge in friction stir welding, 101
Transformation hardening, friction, 46
Traverse speed
   friction stir welding, 101
   friction surfacing, 87, 88
Tunnel defect, 109

U

Ultrasonic additive manufacturing (UAM), 4, 28

V

Vaporization of alloys, 35
VAT photo polymerization, 24
Voids
   friction stir additive manufacturing, 109
   metal additive manufacturing, 34

W

WE53 alloys, magnesium-based, 52, 106, 107, 115
Weight considerations, 2
Welding (friction), 41–44, 59–74, 136–137
   in additive manufacturing technology, 59–74
   advantages, 42–43, 59, 65, 70–72
   angular, 45
   applications/uses, 126–128, 136–137
   definition, 41
   historical developments, 41–42
   linear, *see* linear friction welding
   rotary, *see* rotary friction welding
   variants of techniques, 43–44
Working principles, 19, 45–51
   additive friction stir, 51, 114
   classification of techniques on basis of, 22

Working principles (*Continued*)
   friction-assisted seam welding, 51, 111
   friction stir welding-based additive manufacturing techniques, 98–99, 102–104, 111, 114
   friction surfacing (FS), 45, 84
   joining techniques (in general), 45–51, 60–61, 65–66, 84, 98–99, 102–104, 111, 114; *see also* process

X

X-axis in friction stir welding, 101

Y

Y-axis in friction stir welding, 102

Z

Z-axis in friction stir welding, 101

PGSTL 04/17/2018